Thinking about Thought

*

The Structure of Life and the Meaning of Matter

*

Piero Scaruffi

Volume 1.
Brain

*"Intelligence is not about knowing the answers
but about asking the questions"*

*"What we understand is not enough to understand
why we understand it"*

Scaruffi, Piero
Thinking about Thought - Brain
All Rights Reserved © 2014 by Piero Scaruffi

ISBN-13: 978-1503361065
ISBN-10: 1503361063

(In the USA only one company is authorized to sell ISBNs: Bowker. And Bowker sells them at an outrageous price when other nations issue ISBNs for free. I consider this as, de facto, one of the most blatant scams in any industry. To protest against this government-sanctioned Bowker ISBN monopoly rip-off, I opted to obtain a free ISBN from Amazon CreateSpace, which will then appear as the publisher of this book, and I encourage all authors and publishers to do the same)

Printed and published in the US

For information: www.scaruffi.com

No part of this book may be reproduced or transmitted in any form or by any means, graphic, electronic or mechanical, including photocopying, recording, taping or by any information storage retrieval system, without the written permission of the author (http://www.scaruffi.com)

Contents

PREFACE 4

MIND AND MATTER 11

COGNITION: A GENERAL PROPERTY OF MATTER 46

INSIDE THE BRAIN 67

MEMORY: THE MIND'S GROWTH 111

MACHINE INTELLIGENCE 138

COMMON SENSE: ENGINEERING THE MIND 181

CONNECTIONISM AND NEURAL MACHINES 200

LANGUAGE: MINDS SPEAK 213

THE HISTORY OF LANGUAGE: WHY WE SPEAK 245

METAPHOR: HOW WE SPEAK 272

PRAGMATICS: WHAT WE SPEAK 283

MEANING: A JOURNEY TO THE CENTER OF THE MIND 289

A TIMELINE OF NEUROSCIENCE 307

A TIMELINE OF ARTIFICIAL INTELLIGENCE 311

Preface

By the time you finish reading this book you will be a different person. I am not claiming that this book will change the way you think and act. I am simply referring to the fact that the cells in your body, including the neurons of your brain, are continuously changing. By the time you finish reading this book you will "literally" be a different body and a different brain. Every word that you read is having an effect on the connections between your neurons. And every breath you take is pacing the metabolism of your cells. This book is about what just happened to you.

As with any book worth reading, the goal of this book is to fill a gap. In my case, the gap is a lack of books that provide an interdisciplinary account of the studies on the mind carried out around the world. While many books carry that label, most of them focus on the one or two disciplines or theories that the author intends to defend or attack.

First and foremost, my book aims at providing an accessible and stimulating introduction to those studies across a number of disciplines: Philosophy, Psychology, Computer Science, Mathematics, Biology, Neurology and Physics. This book contains a brief description of every single modern theory (about Consciousness, Cognition and Life) of which I am aware.

This book was originally born to provide an overview for ordinary humans of the philosophical mind-body debate, of neurological models of the brain, of computational theories of cognition (Artificial Intelligence, Connectionism), of post-Darwinian biology, of theories on memory, reasoning, learning, emotions, common sense, dreams, language, metaphor, of modern Physics (Quantum Theory, Non-equilibrium Thermodynamics, Relativity Theory), and, last but not least, consciousness. It was originally meant as a compendium of the scientific ideas that are likely to shape the intellectual scenario of the third millennium. And I was its original reader.

This book also offers a humble personal opinion on what the solution to the mystery of consciousness may be. But that is not the centerpiece of the book.

A popular question of our times is: What is the meaning of life? I always found that question misleading, because first we should be able to answer the more basic question: What is the meaning of matter? This book can't answer either, but at least tries to make the connection.

Physics has explained everything we have found in the universe. We know how the universe started and how it will end. We know what drives it. We know what makes it. Our knowledge of fundamental forces and elementary particles is increasing daily. Two things remain to be explained: how am I alive and how do I think. What does it take for something to be alive and to think? Can we "build" a machine that thinks and is alive? What is thought (consciousness)? And what is life? Physics provides no answer. Historically speaking, Physics never tried to give an answer. Life and thought were considered beyond the reach of formulas. Today, instead, scientists from different disciplines view living and thinking as physical phenomena to be studied the same way we study galaxies and electricity. The most important revolution of our century could be the idea that thinking and living can be explained by mathematical formulas, just like any other phenomenon in the universe. Science may never be the same again, literally. Any scientific theory that does not provide a credible account for consciousness and life is faulted from the beginning, as it ignores the two phenomena its own existence depends upon. We are alive and we are conscious: we know that much.

We live in an age in which the study of consciousness, cognition and life is no longer philosophical speculation. It is, instead, affecting a growing number of disciplines. For the first time in history there is a convergence of specialists (neurologists, biologists, physicists, mathematicians, computer scientists, psychologists) onto one subject.

A new view of nature is slowly emerging, which encompasses both galaxies and neurons, gravitation and life, molecules and emotions. In what represents the culmination of centuries of studying nature, humankind is now approaching the thorniest subject of all: ourselves. We are part of nature, but, historically, Science left us in the background, limiting our role to the one of observers.

For a long time we humans have enjoyed this privileged status. But we seem no longer capable of eluding the fundamental issue: that what we have been studying for all these centuries is but us, albeit disguised under theories of the universe and of elementary particles (theories of what "we" see). And now it is about time that we focus on the real subject. The human mind appears to us as the ultimate and most refined product of life. And life appears to us as the ultimate and most refined product of matter. Now we need a theory of the universe in which consciousness, cognition and life are not oddities, but building blocks.

The fact that we do not have yet a good theory of mind probably means that we do not have a good theory of the universe. Consciousness is perhaps the great mystery of the universe. And the reason may very well lie in a fundamental inadequacy of our Science to explain natural phenomena. In a sense, the new science of mind is doing more than just studying mind: it is indirectly reformulating the program of Science in general.

Future generations will be amazed that it took thousands of years (and hundreds of years since the scientific revolution) to realize how important consciousness is to understand the world and ourselves.

At every point in the history of Science, a paradigm shift allowed for the explanation of previously unexplained phenomena.

The challenge, now, is to explain why we are here.

And what we are.

Ultimately, this book is about the gap between "I" and "me".

This series was begun in 1997. It was published in one volume as "Thinking about Thought" in 2003. A revised and expanded edition titled "The Nature of Consciousness" came out in 2006. This new edition is divided into four smaller volumes and it includes additional material that originally appeared on my website www.scaruffi.com. Click on "Science" and register to the monthly newsletter if you wish to receive future updates.

The website is also the easiest way to find out my email address. I welcome feedback from readers, whether it is typos or opinions.

Piero Scaruffi
Redwood City, 2015

About Me
Few people have the qualifications to write such an ambitious survey of a brand new discipline. I confess i am not one of them. After leading the Olivetti Artificial Intelligence Center for several years, in 1995-96 i spent two years at Stanford University studying Cognitive Science (thanks to a visiting scholarship kindly granted by Robert Engelmore at the Knowledge Systems Laboratory). That was, in retrospect, the beginning of this book. I simply gathered information from all disciplines with the aim of working out a synthesis of sorts. I guess it will remain the goal of my life, although i have missed my chance of working inside academia.

My background is a mess. I graduated in Mathematics but my thesis was on theoretical Physics (that's where i got my introduction to Quantum and Relativity Physics). I worked in the software industry and eventually did research on Artificial Intelligence.

In my other lives, i write about music, cinema and literature, and i am working on a history of knowledge from the beginning of human civilization to our days. I have published poetry that won a few awards. I have traveled to more than 150 countries.

How To Read This Book

I think that all readers will be interested in the main ideas surveyed within each chapter of this series, but probably most readers will not be interested in the details of each and every chapter.

Each chapter contains a short introduction to the subject, and then a series of paragraphs that summarize the theories of the main specialists in the field.

I rarely take sides. I summarize a scholar's work and let you decide.

(I apologize with the scholars: a one-page summary of their work is, of course, a very superficial reading of their theories. The goal of this book is to offer "breadth", not "depth").

What is subjective is the selection: i do make a selection for the reader, not only focusing on theories that represent real paradigm shifts but also omitting (with a few exceptions) those theories that are too unscientific for my taste.

As you advance into the chapter, the theories get more difficult and sometimes repetitive. Depending on your level of interest, you may want to absorb all the details or just skip to the next chapter.

My own ideas are usually left for the end of each chapter. Needless to say, you don't miss much if you skip my ideas.

I have a feeling that, for most readers, the best way to read this book is in many stages: first surf the chapters (focusing on the first half of each chapter), then re-read the book going a bit further within each chapter.

A generous bibliography at the end of each chapter should help you select what you want to read next, depending on what intrigued you most. Titles in bold are those recommended for beginners.

The core material on Cognitive Science (this volume) is divided into chapters that roughly correspond to cognitive faculties (memory, dreams, emotions, language, etc). But the "peripheral" volumes are no less important, and in fact, take up most of the series. The other volumes in the series are devoted to Biology, Linguistics, Machine Intelligence, Physics and, finally, Consciousness. They are mostly modular, except for the sixth and last volume that heavily references the previous five.

Beginning with a survey of Philosophy may not be the best way to introduce a book that promotes a new science. While Philosophy matters less when hard data are available, philosophers of mind did frame the problem. The reader probably has her own strong opinion on what consciousness is and where it comes from. After reading the first chapter and the different theories of all those philosophers, the reader with strong opinions will probably realize that her convictions are a bit amateurish. Thus that first chapter may be helpful to "clear the air".

The volume on Machine Intelligence has a similar purpose. Whether "intelligent" (or, better, conscious) machines are feasible or not, the program of building one has forced people to think about what consciousness is and where it comes from. The fact that we are not even close to building such a machine means that it is either impossible or we are still in the dark.

The volume on Biology provides the kind of general knowledge that is needed to situate cognition in the proper context. Our brain acts in a world. An organism does not live in an empty universe. The origin of life and its evolution are also relevant. The nature of altruism (that seems to defy the essence of Darwinism) is important to understand our mind. The physical and mathematical theories of life are basically abstracting general principles applicable to other fields (possibly neurobiology itself). Throughout these chapters i introduce the basic facts of life (e.g., the genetic code, DNA, the structure of the cell, etc).

Both Neurology and Biology are essential to understand the "what" and the "why" of cognition.

The volume on Language addresses four different dimension (structure, origin, metaphor, pragmatics). The chapter on meaning is rather technical (by the standards of this book) but i felt that a brief introduction to the debate on truth and meaning could not be omitted. Skip it if it sounds tedious.

The volume on Physics is essential. Every other chapter is useful for its philosophical, biological, mathematical or psychological speculations, but, at the end of the day, it is Physics that best summarizes our state of knowledge. Anybody who has an opinion on life, cognition or consciousness without being fluent in Quantum Mechanics and Relativity is simply bluffing.

The last volume attacks consciousness, surveying all the main theories that i am aware of. Its three chapters cover biological theories of consciousness (that are based on classical Physics), theories of how consciousness emerged in evolution, and finally theories of consciousness based on modern Physics (not classical Physics). Like in most of the other chapters, do not be surprised if you find contradictions: there is hardly any consensus yet on any of these theories. Physics is relatively monolithic in its beliefs of what matter is and how it works. No such certainties exist in the science of consciousness. That volume has a separate chapter on theories of the self, revisiting topics addressed in the other volumes.

The character of these volumes varies wildly.

Depending on your background and interests, some chapters may excite you or put you to sleep. Whichever it turns out to be, do not assume that the rest of the chapters are of the same kind. In a sense, each one reflects the style and method of the discipline that it summarizes.

Books that only deal with popular theories of consciousness are not educating the reader: they are merely teasing the reader.

MIND AND MATTER

The Takeover of the Mind

No doubt most people feel that their mind is more important than their body. People may be afraid of losing a limb in an accident, but would still prefer that to losing consciousness. A person who is lying in an irreversible coma is considered "technically dead" even if her body is still alive. We don't mind the transplant of an organ, even of the heart; but we would oppose a transplant of the brain: most people would interpret a heart transplant on them as "somebody is giving me her heart"; but they would interpret a brain transplant on them as "I am giving my body to someone else". We can envision a future in which minds will exist without bodies, but not a future in which we would be happy to be bodies without minds. Ultimately, we are our minds, not our bodies.

It is likely that this was not always the case. There was probably a time when survival of the body was more important than survival of the mind. The preeminence of the mind is a recent phenomenon. The main goal of our ancestors was probably to protect their bodies from predators and from natural catastrophes. If the body dies, the individual (whatever an individual is) simply dies. Nature grants the body an obvious preeminence over the mind, a preeminence that we have forgotten but that was probably there for a long time during the evolution of the human species. For a long time, the human mind may have been simply a means to achieve the goal of protecting the human body. Nothing more than an evolutionary advantage over other species in protecting the body. Just like some animals have a fur to protect them from cold weather. Then, somehow, that evolutionary advantage became the predominant part of the individual. To the point that we declare "dead" somebody whose body is alive but whose mind is not. There has been steady progress towards turning the tables: the mind has slowly taken over the body, and now we think of an individual as her mind (whereas we still think of a dog as its body, regardless of whether it has a mind or not).

Historically, ancient civilizations don't seem to have appreciated how awesome the human mind is, and don't seem to have realized how "low" non-human things are. For example, the ancient Greeks believed that the rivers were children of a god. Today, it may sound strange to think of a river as a "living being", because we know that most of its water changes all the time and we know that its water comes from melting snow and rain, and so forth. But isn't that true of humans too? Don't we change cells all the time? Don't we take in energy and matter from outside (as food)? Doesn't a river have a personality? Other than the fact that rivers live far longer than us, it is not so obvious that having a mind makes humans all that different from rivers, as we today believe.

Historically speaking, the first part of the mystery is why and how minds became more important than bodies. The second part is, in a sense, proof that the mind is a recent accident: we ask ourselves what the mind is (a rather strange question: what am I?). When we ask what the mind is, we implicitly assume that the body is a given. The body is a given and we wonder what the mind is. We don't take the mind for granted and wonder what the body is and why we have bodies. We are bodies that wonder about their minds, not minds that wonder about their bodies. At some point, minds happened to bodies. And now bodies use their minds to wonder "How did that happen" and "What is my mind".

At the same time the body is important in determining who else has a mind. We grant mentally insane and mentally disabled people rights that we don't grant to animals even when those people behave just like animals when those people seem less intelligent than some animals. The main difference is that those people have human bodies. We don't grant any rights to machines, no matter how "smart" they behave, because they don't have human bodies.

The quest for a rational explanation of the human mind started with the task of defining the relationship between mind and matter: is our mind made of matter? Is it made of a different substance? What differentiates the mental from the non-mental? How do our mind and body relate? Is our mind inside our body? Is our mind born with the body? Will it die with the body? Does it grow with the body? These days, having learned quite a bit about the brain and being reassured by countless psychological experiments that our brain is the main entity of the body responsible for our thinking, we are mostly interested in the specific relationship between brain (not "body" in general) and mind: what is the relationship between the neural and the mental? How does the mental originate from the neural?

And, finally, what is "in" the mind?

Dualism And The Mind-Body Debate

Historically, two main schools of thought have antagonized each other: "dualism" and "monism".

According to dualism, mind and body are made of two different substances. The first and most famous of dualists was the French philosopher René Descartes, who is credited with starting the whole "mind-body debate" in 1649. He observed that reality is divided into matter and spirit. These are two different worlds, made of two different substances. He defined what matter is and what mind is: matter is whatever exhibits the property of "extension" (geometric properties such as "size", "shape", etc.) and mind is... "cogito", i.e. thought (a more scientific definition of mind would come later from Franz Brentano). "Res extensa" (things that have an extension) and "res cogitans" (things that think) belong to two separate realms, and cannot be studied with the same tools. This dualism had an enormous influence on future generations. Newton's Physics, for example, is a direct consequence of that approach: Physics studies the realm of matter, and only deals with matter. And such it would remain until the end of the 20th century.

Descartes' dualism was a departure from Aristotle's dualism that had ruled for centuries. Aristotle divided things into living and nonliving. Living beings behaved differently and therefore required a different treatment. Descartes realized that living and nonliving matter are, ultimately, the same matter, that obeys the same physical laws. There is "one" physical world for everything. Living matter appears to "behave" because it is more complex. In reality, animals are mechanical automata. The real distinction is at the level of thought. Some beings (and, for Descartes, it was only humans) can think. The difference is not between living and nonliving matter, which are ultimately the same substance, but between matter and mind, which are two different substances. In a sense, Aristotle's philosophy was centered on life, whereas Descartes' philosophy was centered on the human being. (It would take three centuries to resurrect the idea that animals, too, may have a mind, and therefore return to Aristotle). Descartes also understood that the brain must be the seat of the body-mind interaction, although he couldn't quite explain it.

The 18th century British philosopher David Hume was a dualist too, but he pointed out that "mind" is really a set of "perceptions". The self is an illusion. The mind is simply a theater where perceptions play their part in rapid succession, often intersect and combine. The self is like a republic, whose members have an independent life but are united by a common constitution: the republic is one, even if the members (and maybe even their individual constitutions) are continuously changing. The identity of the republic is provided not by its contents, that are continuously fluctuating, but by the causal relationship that holds its members together.

Epiphenomenalism

The problem with dualism is how mind and body influence each other while being made of two different substances.

In the 18th century the Swiss biologist Charles Bonnet ("Essay on Psychology", 1754) attempted to solve the dilemma by introducing "Epiphenomenalism", the idea that the mind cannot influence the body (an idea later borrowed by the British philosopher Thomas Huxley). Bonnet expanded on Descartes' intuition that mind-body interaction must occur in the brain. He then analyzed the brain and realized that mental activity reflects brain activity. Bonnet also expanded on Descartes' intuition that a body is a mechanical device. He simply added that the automaton is controlled by the brain. Different animals have different functioning (an idea that Huxley married to Darwin's theory) but ultimately they are all bodies run by brains in an optimal way to survive and reproduce. Humans, and possibly other animals as well, are also conscious, but consciousness has no role in directing the automaton. Mind cannot influence the body. The mind merely observes the behavior of the body, although it believes that it actually causes it. (Note that "mind" was pretty much synonymous with "consciousness").

"Epiphenomenalism" therefore accepts that mind and body are made of different substances, but the mind has no influence on the body. The brain causes the mind, but the mind has no saying on the brain's work. Mental events have no material effects, whereas material events may have mental effects. Mental events are simply by-products of material events (like smoke is a by-product of a fire but has no impact on the fire).

The World Of Ideas

Dualists do not doubt that the mind and the brain communicate somehow. But they are faced with the apparently insurmountable task of making two different substances communicate, even though, by definition, those two substances are not supposed to interact. One way out of this dilemma is to assume the existence of an intermediary between the two.

For example, the Austrian philosopher Karl Popper posits the existence of a third world. Since the emergence of language, the human mind has created a whole new world, a world of mental products, which he calls "World 3". Abstract objects of mathematics, scientific theories, artworks and musical compositions are examples of "objects" that belong to neither the mental world (World 2) nor the physical world (World 1). Mind (World 2) plays the role of intermediary between the abstract world (World 3) and the physical world (World 1). Mind (World 2) is basically an operator that related abstract objects and physical ones.

The big difference with Plato's world of ideas is that Popper's World 3 does not preexist humans: it is created by humans, and it evolved according to Darwinian evolution; it has a history. Plato's world of ideas contain concepts, whereas Popper's World 3 contains theories, including tentative theories and theories that might be later proven false. World 3 also contains "conjectures" that affect neural processes of World 1 (a case of "downward causation"). The three world interact with each other in different manners. World 2 has the biological function of relating World 3 to World 1. World 2 can perceive World 3 without any mediation by

World 1. Once learned, a language (World 3 object) has an impact on the personalities of the speakers (World 2 objects). World 3 objects affect World 1 objects through human intervention (World 2 objects).

Predating Richard Dawkins' "memes", Popper argues that cultural evolution is a continuation of genetic evolution by means of World 3 objects.

Plato thought that the mind can intuitively access the world of ideas. Popper thinks that the human mind grasps World 3 objects by a process of reconstruction, a method that is independent of their embodiment. Popper points at the fact that vision (the perception of World 1 objects) is not a passive process but an active one during which the mind "guesses" what object it is seeing, i.e. by formulating hypotheses and then selecting them based on plausibility. That is also how most problem solving is done by mathematicians. Popper imagines that an identical process of "trial and error" is used by World 2 to grasp both objects of World 1 (physical objects) and of World 3 (ideas). Natural selection is not only about the evolution of World 1, but also about the emergence of World 2 (and therefore of World 3, a product of World 2). World 2 and 3 then influenced the evolution of World 1 (or, at least, of those living objects of World 1 that are sentient).

Thus Popper feels that it is incorrect to explain mental life from brain processes (upward causation) and ignore the possibility that mental life drives brain processes (downward causation).

The reductionist view that lower levels determine higher levels (upward causation) but not viceversa is mistaken: levels interact with each other. Popper points out that "downward causation" is at work too: every control mechanism based on feedback determines the behavior of its components, and every tool is a higher-level object that forces each and every constituent molecule to act. The movement of a group of atoms determines the movement of a neighboring group of atoms, i.e. the movement of individual atoms within that group Ecosystems and societies determine the behavior of individuals. The death of an individual leads to the death of all its cells.

Rather than according to hierarchical levels, Popper prefers to see the world organized according to richer and richer domains. Thus Chemistry (that studies molecules) is an enrichment of Physics (that studies the atoms that constitute those molecules). Each "higher" layer may display properties that were not logically implied by the "lower" layer; or, better, new properties "emerge" when a domain is expanded.

Popper thinks that Physics itself, the outcome of the very research of the atomistic project, has proven materialism wrong. Matter is a form of energy, or, better, a process that has to do with transformations of energy.

The universe is not a collection of things but an interacting set of events. Physics has become a theory of matter whose foundations are entities that are not material. As the US physicist John Wheeler put it, one should start the study of matter not with particle physics but with vacuum physics. Another argument against determinism from Physics is that subatomic behavior is largely probabilistic in nature. These arguments lead Popper to believe that unpredictable events do happen, that the universe is "creative". Evolution is one case in which the future was not foreseeable: knowledge of the initial conditions and of the laws of nature was not enough for a sentient being to infer how evolution would proceed, to deduce which species would arise and that one in particular, Homo Sapiens, will become conscious. Atomists/determinists believe that all events are fully predictable, that they are "clocks". Popper contrasts the precisely predictable behavior of clocks with the unpredictable behavior of clouds, concluding that all physical systems, including clocks, are, instead, clouds.

The Australian neurophysiologist John Eccles argues that the interaction between the mind and the brain of an individual is analogous to a probability field of Quantum Mechanics. Mental "energy" can cause neural events by a process analogous to the way a probability field causes action. He calls "psychon" the mental unit that transmits mental intentions to the neural units.

The British physicist Roger Penrose, one of the leaders in General Relativity, also subscribes to the notion that there exists a separate world of conscious states and that the mind can access that world. But Penrose's "world of ideas" is still a physicist's world: "protoconscious" information is encoded in space-time geometry at the fundamental Planck scale, and our mind has access to them (i.e., is conscious) when a particular quantum process occurs in our brain.

A more humble formulation is due to the US mathematician Rudy Rucker, who believes in the existence of a separate "mindscape". Rucker asks: "Is what you thought yesterday still part of your mind?" The question is not easy to answer if you assume that ideas are part of minds. Rucker's conclusion is that there exists a world of ideas separate from the mental and the physical. Our minds can travel the mindscape that contains all possible thoughts just like our bodies can travel the physical space that contains all possible locations. Minds share the same mindscape the way bodies share the same physical space. We all share the same mindscape, just like we all share the same universe. In particular, the mindscape contains all mathematical objects and mathematicians explore mindscape the same way astronauts explore physical space. Ditto for natural laws and physicists. Mathematical formula and laws of nature have an independent existence of their own.

The US philosopher John Searle rejects the idea that the universe can be partitioned into physical and mental properties to start with. After all, things such as "ungrammatical sentences, my ability to ski, the government and points scored in football games" cannot be easily categorized as mental or physical. The traditional "mental versus physical" dichotomy appears to be pointless.

Supervenience

There exist two main brands of dualism: "substance" dualism (the mind is a different substance altogether from the brain), and "property" dualism. According to the latter, the mind is the same substance as the brain, but comes from a class of properties that are exclusive of the brain. The main version of property dualism is "supervenience" theory.

"Supervenience" is used to express the fact that a domain is fully determined by another domain. For example, biological properties "supervene" (or "are supervenient") on physical properties, because the biological properties of a system are determined by its physical properties. Biological and physical properties of an organism are different sets of properties, but the physical ones determine the biological ones. Nonetheless, one can study only the biological properties and never deal with the physical ones.

The Korean-born philosopher Jaegwon Kim ("Concepts of Supervenience," 1984) applied the concept to mind: mental properties are supervenient on physical (neural) properties. According to Kim, then, the mental is supervenient on the physical just like the macroscopic properties of objects supervene on their microscopic structures. Intuitively this means that mind is to brain what lightning is to electrically charged particles: the same phenomenon, that presents itself in two different ways.

Kim's supervenience defines a relationship between mental and physical, and it also defines some constraints. A mental state cannot correspond to two different physical states. Two brains can't be in the same mental state and be in different physical states. Mental states depend on corresponding neural states: any change in mental states must be matched by a corresponding change in physical states. Mental states "are" neural states, the same way that electricity "is" electrons.

One can organize nature in a hierarchy, starting with elementary particles and ending with consciousness. At each level some properties apply, but at the immediately higher level some other properties apply. For example, electrons have mass and spin, but electricity has potential and intensity. Chemical compounds have density and conductivity, whereas biological organisms have growth and reproduction. At each level a new set of properties "emerge": the weak force at the elementary particle level, viscosity at the molecular level, metabolism at the biological level, and so forth; and consciousness at the cognitive level.

The British philosopher Charles Dunbar Broad had already showed in the 1920s that the universe is inherently layered and that each layer yields the following layer but cannot explain the new properties that emerge with it.

Supervenience takes it for granted that nature works this way, but offers no explanation of why, at a higher level, we would find electricity instead of, say, "huicity" or "flowixity" (imaginary properties): why and how just those properties? Why and how does the mind emerge from the brain? Ultimately, this is the dilemma of "mental causation": how does the brain cause the mind? In general, this is the dilemma of "second-order properties": how do properties at one level cause properties at another level?

John Searle (who believes that minds are high-level features of brains) admits supervenience to the extent that it is causal: the same neural states are also the same mental states because the former cause the latter. Searle thus reduces supervenience to causality. But Kim does not impose any causal relationship: the relationship between the mental and the neural is analogous to the relationship between the usefulness of an object and the features that make it useful: those features do not "cause" its usefulness, they "constitute" its usefulness.

All facts of the universe depend (and are therefore supervenient) on physical facts, but the nature of such "dependence" is not trivial, according to the Australian philosopher David Chalmers. Properties that are supervenient on the physical world can normally be reduced to it (i.e., explained in terms of it), but consciousness is not truly, completely supervenient on the neural, and therefore it cannot be reduced to the neural. Consciousness is to some extent supervenient on the physical, but (by the nature of its kind of supervenience) it cannot be explained in physical terms.

Monism

There is an obvious alternative to dualism: monism. According to monism, body and mind (matter and thought) are made of the same substance: "idealists" think that everything is mental, "materialists" think that everything is material. So monism mainly divides into idealism and materialism.

But the "one" substance that everything is made of can also be something else than matter or mind.

The Dutch philosopher Baruch Spinoza (17th century), for example, believed that only one substance exists, which is both infinite and eternal, and that "the" substance is conscious and it has extension. This substance is expressed in an infinite series of "modes". Humans only perceive two of those modes because we are equipped with only two attributes of that substance, hence we see a world of minds and bodies. When we perceive modes through the attribute of thought, we perceive ideas, and we perceive them through the attribute of extension, we perceive objects. God is all that exists (he is what is), there is nothing that is not God.

The British philosopher Bertrand Russell was also a monist of sort, because he believed that everything in the universe is made of spacetime events which are neither mental nor physical. His "neutral monism" reprises ideas from the Austrian physicist Ernst Mach and from the US psychologist William James ("Does 'Consciousness' Exist?", 1904): there is a fundamental constitutent of the universe which is neither mental nor physical but yields both the mental and the physical that we observe.

Idealism

According to idealism, mind is the only substance that makes up all of reality.

The German philosopher Gottfried Leibniz (17th century) believed that only minds exist. Humans are not the only ones to have minds. Everything has a mind. Even matter is made of minds. Minds come in degrees, starting with matter (whose minds are very simple) and ending with God (whose mind is infinite). Reality is the set of all finite minds (or "monads") that God has created. Everything has a mind. This extreme view of idealism is called "panpsychism". One way to get rid of the mind-body problem was to get rid of the body.

The Irish philosopher George Berkeley (18th century) thought that all we know is our perceptions, and whatever concepts we can build up from them ("esse est percipi"). We cannot directly know that there is an external world. We only know the internal world of our perceptions. When we talk of an object, we talk of what we see, hear, taste, touch, smell: we talk of something that is inside our mind. An object is an experience. The whole universe is a set of experiences. Ultimately, the only thing that exists is the experiences of our mind.

In the 1920s the British mathematicians and philosopher Alfred Whitehead proposed that mental life occurs in a field of protoconscious events. His units are similar to Leibniz's monads, but they are limited in time, and therefore better thought of as "mental events". Mental life is a sequence of such mental events that occur in this mental space.

As brain studies have proved that the senses present us with a fictitious view of the universe, and subatomic Physics has shown that matter is but clouds of floating particles, and Quantum Mechanics has stated that reality is ultimately in the observer's mind, it has become more tempting to embrace idealism. If everything we see and hear is but an illusion, how can we claim that there really are "things" out there? The only thing that we perceive is what the senses fabricate for us. What we call "reality" is the work of our mind. If Physics even predicts that reality cannot be "measured" without an observer (as Quantum Mechanics does), how can we claim that reality exists independent of our mind?

The problem with idealism is that one cannot do much more than claim to be an idealist. Once that claim has been made, reality cannot be used to prove it, since reality is a mere illusion of our mind. Everything is an illusion, including the things that one could use to prove this statement right or wrong.

Most scientists believe in a milder form of idealism: the senses do fake reality, and reality does need an observer to become what it is, but sensations do relate to an external world and measurements do measure an external world. The senses and the brain simply alter reality so that we can move about and survive in a world that we can comprehend and manage. And Quantum Physics does not forbid reality from existing, it only forbids us from completely perceiving it.

Most scientists believe that the reality that we perceive is indeed a fabrication of our mind, but it does correspond to a reality out there, that exists regardless of our mind.

Neutral Monism

Philosophers have been debating for centuries whether there are two substances or there is only one substance, whether dualism or monism are the right model for the world. Bertrand Russell reached the conclusion that, if there is a substance, it is neither mental nor material, but, best of all, is to assume that there is no substance at all. His ideas have been largely neglected, which is surprising since his ideas are the only ones in the entire philosophy of mind to be truly based on an understanding of Physics.

Russell simply took Einstein literally: if space and time are inseparable, if matter is energy, if everything is relative to the observer, then both matter and mind are meaningless oversimplifications of reality. Matter is less material than Newton thought, and the spirit is less spiritual than Berkeley thought. Neither truly exists as a substance. They are rather different ways of organizing space-time. What truly exists is "events". I am a cluster of space-time events that sticks together for a little while.

The same argument can be seen from the viewpoint of perception.

Sensations are both material and mental. A sensation is part of the object that can be constructed out of it. A sensation is part of the mind in whose biography the perception occurred. An object is defined by all the appearances that emanate from the place where it is towards minds. A mind is defined by all the appearances that start from objects and reach it. If we represent the universe as a network of interactions between many objects and many minds, an object is the collection of all its outputs, a mind is the collection of all its inputs. An object is not the generator of such outputs and a mind is not the receiver of such inputs.

The difference between matter and mind is simply the "causal" relationships that are brought to bear.

There is no substantial difference between matter and mind. They are built out of the same stuff, which is neither material nor mental (it is "neutral").

Materialism

According to materialism, instead, mind and body are made of the same substance, which is matter, as defined by Physics; and the mental can be explained from the physical.

This position was first embraced enthusiastically in the 18th century by the French philosopher Julien Offroy de la Mettrie who envisioned the "Homme Machine", the mind as a machine made of matter, and thought as a material process. Unlike dualism, materialism, in all its variants, admits only one kind of substance, and one class of properties.

Materialism had its golden age following a paper published by the Austrian philosopher Herbert Feigl ("The Mental and the Physical", 1958). It was his paper that established the "mind-body problem" at the center of 20th century philosophy, after so many decades of neglect.

Dualism and materialism have been the protagonist of the centuries-old mind-body debate. They both have their pluses and minuses, and neither can overcome its minuses in a plausible way. Dualism's plus is that it does recognize the difference between conscious and non-conscious matter; its minus is that it cannot explain how the mind and the brain connect. Materialism's main asset is that it does not need to explain that connection, since the mind "is" the brain; its drawback is that it cannot explain how consciousness arises from non-conscious matter.

Behaviorism

A position that tried to get rid of the mind versus matter debate is "behaviorism", which deals with mental terms (such as "belief", "hope" and "fear") only to the extent that they are related to behavior.

Following the lead from the US psychologist John Watson, who had already ruled the mental out ("Psychology as the Behaviorist Views it", 1913), behaviorists reject mental states as unscientific, as an annoyance of our language. What matters is only the relationship between disposition to behavior and actual behavior.

In particular, the British philosopher Gilbert Ryle argued that the mind is not another substance but simply a domain of discourse. Ryle took issue with words that refer to mental objects as if they were on the same level as physical objects. In his opinion, they are not. They are merely words used to describe "behavior". The mental vocabulary does not refer to the structure of something, but simply to the way somebody behaves or will behave. The mind "is" the behavior of the body. Physical objects exist, mental objects are merely vocabulary.

Descartes invented a myth: the myth of the mind inside the body. A myth which Ryle parodied with the famous expression "the ghost in the machine": we assume that we have a mind, we ascribe a life to our mind, and when we can't find mind in nature we decide that mind is a different substance or property. Behaviorism is not interested in discussing the mind, but simply behavior and disposition to behavior. Sentences about mental states become scientific and meaningful only when they are translated into sentences about actual and possible behavior. For a behaviorist, a person in pain is simply a person who cries and does some other things that we associate with the word "pain".

Psychological behaviorism went even further in claiming that all behavior can be explained as stimulus and response relations. Behaviorism therefore rejects the common-sense notion that our mental states cause behavior. That is just an illusion.

Behaviorism was briefly popular but the renewed fight between dualists and materialists quickly eclipsed it.

The Age Of Materialism
As the mind-body problem became the "mind-brain problem" during the course of the 20th century, materialism begot "physicalism", according to which a mental state "is" a physical state of the brain.

Note that the emphasis on the brain is not completely natural: I feel pain in my foot, not in my brain. But progress in neurophysiology has created a fascination with the brain, which some people describe as the most complex thing in the universe. Therefore the emphasis shifted from the body to the brain although there really is no evidence to rule out that the rest of the body does not affect the mind (there is evidence to the contrary).

The "identity theory" (the one pioneered by Feigl) states, point blank, that mental states "are" physical states of the brain (just like lightning is "identical" to a stream of electrons). Since it is a little implausible to assume that for every single mental state there is a unique neural state, a variant of identity theory relaxed this constraint: the "token identity theory" (Donald Davidson) does assume that any mental state is identical to a physical state, but the physical state corresponding to a given mental state is not necessarily always the same one (this allows for two people to have the same feeling, or for the same person to have the same feeling twice, without having every single neuron in the same state both times).

The most difficult problem for materialism is to explain how the mind, and especially feelings, can arise from material processes: how can electrochemical activities in my brain suddenly turn into the feeling of pain or fear? John Searle has summarized the problem as a paradox: either the identity theory leaves out the mind, in which case it is implausible, or it does not, in which case it is not materialist anymore.

The Identity Theory

The main issue with any materialistic theory is how the mind (thoughts, feelings) can be explained from what we know of matter. If mind is ultimately matter, then what is it made of and how is it built? How, in other words, can the mental be reduced to the physical?

The "Identity theory" was first advanced by the British (albeit Australian residents) philosophers Ullin Place ("Is consciousness a brain process", 1956) and John Smart ("Sensations and brain processes", 1959). They claimed that perceptions and consciousness are physical processes in the brain, just like lightning is a physical process in the air. They went as far as identifying conscious states with brain states. This removes the question of where the mind-body interaction occurs: since conscious states and physical states are the same thing, they don't need to interact. They "behave" together. A desire, for example, is both a conscious and physical state that causes some actions that are both conscious and physical states.

As Herbert Feigl put it, mental states and physical states have the same "extension" but different "intension": they describe the same states, but in a different way. Mental idioms and physical idioms are different descriptions of the same states. From the viewpoint of the man on the street, this thesis is difficult to defend, as mental and physical states are "obviously" different (pain, fear, love as opposed to electrochemical processes in the brain). An old philosophical trick, the so called "Leibniz's law", holds that two things are identical if and only if all the properties that apply to the first one also apply to the second, and viceversa. But the properties of mental states (such as emotions) and the properties of physical states (such as electrical and chemical properties) are obviously different.

There are several variants of the identity theory.

Instead of limiting the identity theory to consciousness and sensations, the US philosopher David Lewis ("An argument for the identity theory", 1966) and the Australian philosopher David Malet Armstrong extended it to everything that is mental, not just consciousness and perceptions: all mental states are physical states in the brain, all mental processes are brain processes. Furthermore, mental states have a "causal" role: a mental state (eg, a belief, a desire, a fear) may cause behavior, and it does so because it is a brain state. A mental state (which is a brain state) both is caused by and causes behavior. For example, lightning is not only a physical process in the air: it is caused by that physical process. A mental state is defined by its causal role: what causes the mental state, what behavior the mental state in turn causes, and its relationship with other mental states.

Whatever their spin on the identity theory, all these philosophers faced the same problem: how to explain the emotions we feel, which are obviously very different in nature from a piece of matter.

Against Physicalism

There are two main arguments against physicalism. The "knowledge argument", by the Australian philosopher Frank Jackson ("Epiphenomenal Qualia" 1982) is about a scientist who has a complete understanding of the science of color, but has never experienced color: will she learn something new the first time that she experiences color? If yes, then it means that there cannot be a complete physical explanation of mental states.

The zombie argument was originally proposed by the US philosopher Saul Kripke ("Naming and Necessity", 1972) using the "philosophical zombie" thought experiment by the US philosopher Thomas Nagel ("Armstrong on the Mind", 1970). If a world in which all physical facts are the same as those of the real world must contain everything that exists in the real world, then a world of non-conscious (zombie) human beings physically identical to the real world of conscious human beings must contain consciousness. Along those lines, David Chalmers thinks that p-zombies prove physicalism wrong: if we can imagine a world of p-zombies that is identical to ours, then consciousness is not only due to physical processes. Daniel Dennett, instead, counters that a world of p-zombies is logically impossible, that the very act of imagining a zombie is impossible and it leads to contradictions, just like imagining a functioning body with no health.

The Irreducibility Problem

One elegant way of solving the "irreducibility" problem (how mind can be reduced to matter) was devised by the British philosopher Bertrand Russell. He was keenly aware of the inscrutability of matter in general and of brain matter in particular: we cannot know the nature of matter (electrons, gravitational waves and so forth) other than through theories and experiments, but never feel it directly. In particular, we cannot know the processes that occur in our own brain.

Russell thought that mind allows us to perceive, at least, some of those processes as they occur in the brain. He made the remark that what a neurologist really sees while examining someone else's brain is a part of her own (the neurologist's) brain. The irreducibility of the mental to the physical is simply an illusion: the mental and the physical are different ways of knowing the same thing, the former by consciousness and the latter by the senses. Consciousness gives us immediate, direct knowledge of what is in the brain, whereas the senses can observe (possibly aided by instruments) what is in the brain.

In Russell's theory, the mental is not reduced to the physical, and the traditional preeminence of the physical over the mental is turned on its head: the mental is a transparent grasp of the intrinsic character of the brain. Consciousness is, basically, just another sense, a sense that, instead of perceiving colors or smells or sounds, perceives the very nature of the brain.

Anomalous Monism

Donald Davidson ("Mental Events", 1970) showed that the mental and the neural are not the same thing while avoiding Descartes' dualism and remaining within the scope of materialism.

Davidson's theory of the mind rests on a simple syllogism:
- At least some mental events interact causally with physical events
- Events related as cause and effect usually fall under strict deterministic laws
- But there are no strict deterministic laws under which mental events can be predicted and explained (this is the "anomalism" of the mind).

This reads like a contradiction, unless we assume that the mind is something else. What this all means is that the physical and the mental realms have essential features that are somehow mutually incompatible: a mental state cannot just be a brain state. There can be no laws connecting the mental with the physical. In other terms, there can be no theory connecting Psychology and Neurophysiology.

In the lingo of identity theory, Davidson claims that the same instance of a mental state may correspond to different neural states at different times; which means that, given a mental state, it is not possible to relate it to a specific physical state. The same event is both mental and physical, but there is no formal relationship between the two descriptions.

Every mental event is a physical event, but it is not possible to reduce mental properties to physical properties (there exist no "psychophysical" laws), and therefore, for example, the language of Psychology cannot be reduced to the language of Physics. The mental is ultimately physical, but there is no way to explain a mental event in terms of physical events. The mental domain cannot be the object of scientific investigation. (Ultimately, we can view this as due to the fact that the mental is holistic).

Donaldson's "token identity theory" came to identify a less strict version of identity theory: more than one physical state may correspond to the same mental state (a mental state can be realized in several different physical states). This would account for the fact that people with widely different brains can be in the same psychological state.

This view opened the doors to functionalism.

The Age Of Functionalism

If a mental state can be realized in more than one physical state, is the physical state important at all?

What is it that makes a physical state of the brain also a mental state? "Functionalists" had an answer: it's the "function" it performs in the life of the organism. This function will cause a behavior.

The physical state is not important in determining a mental state, but the function is. We call something a "thermometer" if it measures temperature, regardless of whether it is made of plastic or metal: it is the function, not the material, that determines what things are. Likewise, a mind is a mind if it has the function that a mind has, and it doesn't really matter what it is made of.

The "function" of something is a combination of the stimuli it processes, the operation it performs and the external behavior it causes.

Functionalism (basically introduced by Armstrong and Lewis) is really a special case of token-identity materialism in which a mental state is defined uniquely by the causal relation it bears over behavior and over other mental states. Mental states express, ultimately, causal relations (the occurring of something causes something else to take place). In other words, they ultimately have a function. Never mind what they are made of: mental states have a function and that is what matters.

A consequence of this approach is that a mind doesn't necessarily require a brain: anything that can play the same function is a mind too. Mental states are defined by their function, and they may as well be implemented in a computer or a brain. As a matter of fact, by using a technique inspired by the British mathematician Frank Ramsey, it is possible to translate every sentence containing "unscientific" psychological terms (such as "believe", "desire", etc.) into a more formal sentence (akin to sentences of predicate logic) which only contains causal relations. The mind is simply the location at which these causal relations are carried out.

Of course, the difference between functionalists and behaviorists is not so clear cut. Basically, behaviorists refused to deal with mental states and focused on behavior, whereas functionalists said that mental states are such because they cause behavior. Functionalism does not deny the existence of mind, actually it extends the possible realizations of mind in nature (it doesn't have to be a brain).

Functionalism has an advantage over materialism: there is evidence that different neural circuits cause the same mental states (different people with different brains feel the same emotions, the same person with a changing brain feels the same emotions, a damaged brain tends to repair itself to perform the same chores it was doing before), but materialism entails that a mental state is a direct consequence of a physical state, which could be meant to signify that two different physical states yield two different mental states. Functionalism allows for "multiple realization". Strictly speaking, it doesn't even require that the mental state be realized in a brain: functionalism is only concerned with the "function", not with the thing that performs the function.

Since functions must be carried out by a physical entity, functionalism implied some kind of materialism. David Lewis explicitly married the two, materialism and functionalism: every mental state is a physical state, and every mental state is a functional state. The marriage of the two solved two classes of popular paradoxes, the "mad pain" paradox (what if a human was born who is made of flesh like everybody else but his reaction to the feeling of pain is completely different?) and the "android pain" paradox (what if a being made of different stuff reacted to pain the same way humans do?)

The whole debate on one or two substances is meaningless: what is relevant is the function, not the substance. A mind can be "implemented" in whatever number of substances, as long as it performs the function of a mind.

Computational Functionalism

But how do mental states cause physical behavior? This was still the old conundrum of dualism: how do mind and body interact?

A possible solution was found by analogy with a device that had become very popular in the 1950s: the computer. The computer implemented the very concept that a substance (the software) can influence another substance (the hardware).

Functionalism thus begot "computational functionalism" (Hilary Putnam, Jerry Fodor, Stephen Stich, Ned Block), according to which the mind is a program and the brain is its hardware, and the execution of that program in that hardware yields a result which is the external behavior of the organism.

The mind is a symbol processor (just like a computer) and mental states are related to computational states.

Another special case of computational functionalism is "homuncular" functionalism (Daniel Dennett, William Lycan), which decomposes the mind into smaller and smaller minds until it reduces to a physical state: a mental process is the product of many lower mental processes, and each of these lower processes is the product of more and more primitive processes. Each lower layer is less "mental" than the previous one. At the bottom of this hierarchy are the neural processes of the brain.

The most common critique of functionalism is that it is utterly implausible that objects different from a brain can have a mind. But then (as Chalmers has pointed out) the brain itself, that ugly, messy, sticky mass of gray and white matter, is an unlikely candidate for something so special as a mind. Why should a computer look more bizarre than a brain?

Does mind reside in organization or in substance? Or both?

The Computational Theory of the Mind

The US philosopher Hilary Putnam ("Minds and Machines", 1960) focused on the fact that the same mental state may be implemented by different physical states. For example, each person has a different brain, but every person has the same psychological states of "fear", "happiness", etc. Even other animals exhibit some of the same states. Putnam classified mental states based on their function, i.e. their causal roles within the mental system, regardless of their physical structure. Physical states and mental states can even be grouped in different ways.

Putnam then suggested that the psychological state of an individual should be identified with the state of a so-called "Turing machine" (basically, with a computer). A psychological state would cause other psychological states according to the machine's operations. Belief and desire would correspond to formulas stored in two registers of the machine. Appropriate algorithms would process those contents to produce action.

Putnam's idea led to a special case of identity theories, the "computational theory of the mind".

The "representational theory of the mind", developed by the US linguist Jerry Fodor, is an evolution of Putnam's ideas. Fodor argues that the mind is a symbolic processor. Knowledge of the world is embedded in mental representations, and mental representations are symbols, which possess their causal role in virtue of their syntactic properties (i.e., in virtue of how they can be used in "computing" operations). The mind is endowed with a set of rules to operate on such representations. Cognitive life is the transformation of those rules. The mind processes symbols without knowing what those symbols mean, in a purely syntactic fashion. Behavior is due only to the internal syntactic structures of the mind.

The symbols used to build mental representations belong to a language of thought, or "mentalese". Such language cannot be one of the languages we speak because the very ability to speak requires the existence of an internal language of representation. Such language is an intrinsic part of the brain and has been somehow produced through evolution. A belief, for example, is realized as a sentence in the language of thought which resides in the belief area of the brain ("I believe that my name is Piero" is implemented in the belief area by the translation in the language of thought of the English sentence "My name is Piero").

This inner language of thought is shared by all creatures capable of "propositional attitudes" (the simplest form of thought, such as beliefs, hopes, fears, desires). Such creatures can then express their representations in whichever human or animal language they happen to speak.

Fodor basically offers a solution to the problem faced by dualists: how to connect the mind and the body, mental states and physical states, the desire to do something and the act of doing it. Beliefs and desires are information, represented by symbols, and symbols are physical states of a processor, and the processor is connected to the muscles of the body. When the symbols change, they have an impact on the body, they cause behavior. At the same time, perception results in a change of those symbols. The processor, in turn, may change the symbols because it compacts several of them into a new one (reasoning). Mind and body communicate via symbol processing.

Fodor's computational theory is consistent with those offered by the US linguist Noam Chomsky in linguistics and later by the British psychologist David Marr in vision: the mind as a set of modules that "compute" something based on an innate symbolic capability. Noam Chomsky spoke of "mental organs", to relate their role to the role of physical organs. Each organ carries out a function and communicates the results to the other organs.

Fodor generalizes their ideas: the mind is made of genetically-specified modules, each one specialized in performing one task. A module corresponds to a physical region of the brain, and is isolated from other modules. A module receives input only from modules of lower level, never from higher levels (for example, a belief cannot influence the working of a module that analyzes sensory data). Each module generates output in a common format, the "language of thought". Their outputs are input to the central processor, that manages long-term memory and manufactures beliefs. The central processor is the only module that is not domain-specific. Every other module deals with a specific domain.

Fodor does not seem to contemplate cognitive growth: the modules are fixed at birth and remain the same throughout the life of the individual.

The approach of the Canadian philosopher Stephen Stich is even more purely syntactic: he even rejects the notion that each object of a mental operation must represent something (or stand for something). Stich assumes that cognitive states correspond to syntactic states in such a way that causal relationships between syntactic states (or between syntactic states and stimuli and actions) correspond to syntactic relationships of corresponding syntactic objects. His "mind" is a purely syntactic program.

The US philosopher Ned Block believes that the psychological state of a person can be identified with the physical process that is taking place in the brain rather than the state in which the brain is. A psychological state can then be represented as an operation performed on a machine, i.e. identified with the computational state of the machine, rather than with its physical state. This way the psychological state does not depend on the physical state of the machine and can be the same for different machines that are in different physical states, but in which the same process is occurring.

Block has, actually, provided the broadest criticism of functionalism ("Troubles with Functionalism", 1978). "Qualia" (sensations that are associated to the fact of being in a given psychological state) are not easily explained in a functionalist view. Take an organism whose functional states are identical to ours, but in which pain causes the sensation that we associate to pleasure ("inverted qualia"), and an organism whose functional states are identical to ours, but in which pain causes no sensation ("absent qualia"): functionalism cannot, apparently, account for either case. Furthermore, functionalism does not prescribe how we can limit the universe of organisms who have mental states. A functionalist might think that even Bolivia's economy, as expertly manipulated by a financier, has mental states.

Homunculi

The basic picture of homuncular functionalism was given by the US philosopher Daniel Dennett. One can explain the feelings of mental life from the non-conscious behavior of the physical world by an infinite regression: a mind is made of a number of simpler minds that are made of even simpler minds, and so forth. We eventually reach levels at which the minds are performing very simple tasks, like computing 1+1 or deciding whether a color is dark or light. At that level it is not difficult to accept the idea that those elementary "minds" are physical, non-mental, things, such as brain processes. Just like adding infinitely-small instants eventually yields a finite interval of time, so adding "infinite" levels of "homunculi" eventually yields a mind.

Dennett reduces the mind to a set of cognitive functions, and then each function to simpler cognitive tasks, each time reducing the "intelligence" needed to solve the problem. Eventually this process reaches a level at which problems can be solved with no more intelligence than the one that can be found in a mechanical device.

The idea is that, at each level in the organization of a system, the overall behavior of that level is given by the interaction of a set of interconnected components ("homunculi"). The behavior of each component is itself defined by a set of interconnected components at the lower level.

Another US philosopher, William Lycan, thinks that, besides the low level of electrochemical processes and the high level of psycho-functional processes, nature is organized in a number of hierarchical levels (subatomic, atomic, molecular, cellular, biological, psychological). And each level is both physical and functional: physical with respect to its immediately higher level and functional with respect to its immediately lower level. Proceeding from lower levels to higher levels we obtain a physical, structural, description of nature (atoms make molecules that make cells that make organs that make bodies...). Proceeding backwards, we obtain a functional description (the behavior of something is explained by the behavior of its parts). The "aggregative ontology" ("bottom-up") and the "structured epistemology" ("top-down") of nature are dual aspects of the same thing. The apparent irreducibility of the mental is due to the irreducibility of the various levels.

In a similar vein to Dennett's homunculi, the theory of the "society of mind" advanced by the US computer scientist Marvin Minsky assumes that intelligent behavior is due to the non-intelligent behavior of a very high number of agents that are organized in a bureaucratic hierarchy. The set of their elementary actions and their communications can produce more and more complex behavior, and eventually mental life as we know it.

Eliminative Materialism

"Eliminative materialism" is the doctrine that mental states do not exist, or, at least, that the terminology of the mental is wrong and should be abandoned.

The first one to point out that the mental vocabulary constitutes a sort of "folk psychology" was probably the US philosopher Wilfred Sellars

("Empiricism and the Philosophy of Mind", 1956). That theory, that has not progressed for millennia, mostly fails to explain and predict behavior, and it is founded on knowledge about human beings that has long been proven false. Folk psychology is not a science.

The German philosopher Paul Feyerabend ("Mental events and the brain", 1963) and the US philosopher Richard Rorty ("Mind-body identity", 1965) denied the existence of the mental. They claimed that sensations are not brain processes, but the things that we think are sensations are indeed brain processes. The mental is nothing more than a myth. As the US neuroscientist Paul Churchland argues that our introspection cannot be trusted as our other senses (such as sight and hearing) mislead us about the real structure of the universe: for example, we don't see or hear elementary particles and probability waves. It is only the vocabulary of our "folk psychology" that refers to beliefs and desires, sensations, emotions, thoughts, etc. We explain people's behavior by using this terminology, which ascribes mental states to people. In reality, only brain processes exist. In his opinion we should replace the outdated language of folk psychology with the language of neurobiology, just like folk physics was replaced by the more precise language of Newton's physics. Terms such as "belief" and "desire" are as scientific as the four spirits of alchemy.

Churchland points out evidence that folk psychology is unscientific: 1. it has remained the same since the ancient Greeks (but so did arithmetic, didn't it?); 2. it does not integrate well with the natural sciences (but it has been integrated with computer science by computational functionalism); 3. it is incomplete, as its vocabulary does not apply well to mental phenomena such as sleep and mental diseases (Newton's physics was also incomplete, but that does not mean that the terminology of mass and energy should be abolished).

Churchland denies any validity to "first person" mental life, to consciousness, the self, emotions, etc. He grounds his objection to the fact that there is nothing in the brain that resembles what folk psychology talks about: there are only patterns of neural activity.

Intentionality

In their search for "the" ultimate definition of what is the mind, for the one property that differentiates the mind from anything else, a recurring popular candidate has been what philosophers call "intentionality" (from the Latin "intendo", which means "to point at"). "Intentionality", or in-existence, of an object is a concept originally introduced by the medieval Scholastics. Their "intentionality" bears no relationship whatsoever to the modern English word "intentional". Their intentionality is the property of referring to something else.

Mental states have the (apparently) unique property of referring to something else. For example, we are afraid "of" something, we believe "in" something, we know something. Intentionality is the property of being "about" something. "Fearing", "knowing", "believing" are intentional states. If no other natural phenomena exhibit intentionality, then intentionality could be assumed to be the feature that differentiates the mind from the rest of natural phenomena.

All this was summarized in 1874 by the Austrian philosopher Franz Brentano in his influential "thesis": all mental phenomena are intentional; no physical phenomenon is intentional; therefore mental phenomena cannot be reduced to physical phenomena; intentionality is what sets apart mental and physical systems.

Brentano noted that every mental phenomenon includes something as an object within itself, although the way it is included is not always the same (in love something is loved, in hate something is hated, and so forth). "This intentional in-existence is characteristic exclusively of mental phenomena." Every thought we have is about something: we love, we hate, we believe in, we fear that, we hope that... something.

Intentionality comes in different "flavors", also known as "propositional attitudes", and later philosophers focused on four basic ones (belief, desire, hope, know).

Brentano's disciple Alexius Meinong (in his 1904 "theory of objects") even went as far as to state that mental states must have their own existence apart from the physical world. My belief in something is realized by a mental state of something that exists although in a different form than the one in which physical objects exist.

What Brentano said was that all mental states are "representations" of objects. What Meinong said was that those representations exist apart from the objects they represent.

Neither Brentano nor Meinong explained how these "representations" are generated and what they are made of.

There can be many consequences stemming from the theory that the mental is intentional. Brentano's conclusion from his thesis was dualism: the mental and the physical are different substances, and intentionality helps us to discriminate them and study the mental.

More than half a century later (in 1960) the US mathematician and philosopher Willard Quine reached a different conclusion: that intentionality is meaningless, because it does not relate to anything physical. Jerry Fodor believes that the mental is intentional, but it can be reduced to the physical. Daniel Dennett thinks that intentionality is simply a "stance", one of the many we can adopt in studying a system. And not everybody agrees that the intentional can only be mental: the US philosopher Fred Dretske has studied it as a general property of systems. Any device that carries information exhibits some degree of intentionality.

Intentionality and consciousness are the key features of the mind. What is the relationship between them? John Searle says that everything that is intentional is either conscious or potentially conscious. Intentionality would then be an "enabling" feature of consciousness. A system would be intentional before it could be conscious. Still, what makes an intentional feature conscious? Why is intentionality a prerequisite to consciousness?

The Intentional Stance

Daniel Dennett thinks that the folk concepts of belief and desire define a fundamental aspect of our language: they help us explain the behavior of systems (including ourselves). But he denies that they have any physical existence of their own.

In order to explain and predict the behavior of a system one can employ three strategies: a "physical stance", which infers the behavior from the physical structure and the laws of Physics; a "design stance", which infers the behavior from the function for which it was designed (we know when a clock alarm will go on even if we don't know the internal structure of the clock); and an "intentional stance", which infers the behavior from the beliefs and desires that the system must exhibit to be rational. These are simply three different ways of speaking about the same thing. They are more like three different vocabularies, or languages, than three different sets of things. The "intentional stance" is therefore only a particular way of speaking about systems in general, and our mind in particular.

"The tree needs water", "The car wants to be washed", and so forth are examples of the "intentional stance". It is another way to describe the state of objects: the intentional stance. The "intentional stance" is merely the set of beliefs and desires of an organism that allow an observer to predict its actions. Belief and desires are not internal states of the mind which cause behavior, but simply tools which are useful to predict the behavior. No system is truly intentional.

The beliefs and desires of an organism, and how they affect the organism's behavior, have biological origins. If an organism survived natural selection, the majority of its beliefs are true and the majority of its desires are possible, and the way the organism employs them is the most "rational" (beliefs are used to satisfy the organism's desires). If this were not the case, the organism would not have survived. Intentional systems are rational systems, by definition. Intentionality and rationality are complementary aspects of natural selection. The fact that some intentional systems are also cognitive systems is a detail. Beliefs and desires are, first and foremost, biological products, and have a biological function. Ultimately, these intentional stances are but descriptions of the relationship between the intentional system (e.g., the mind) and its environment. An entire organism can be described by its intentional stance, since it is a product of natural selection.

Facts described from the intentional stance can be explained from the design stance, which is in turn grounded in the physical stance.

Mind is what we ascribe to objects, including other humans and ourselves, when we use the intentional stance.

It is not a different kind of matter or a different kind of property, it is just a way of describing what happens.

Intentionality as Representation

Fred Dretske's theory of intentionality resorts to Claude Shannon's and Warren Weaver's theory of information: a state transports information about another state to the extent that it depends on that other state. By the same token, intentionality can be reduced to a cause-effect relationship: each effect refers to its cause. From Dretske's perspective, the intentional idiom of beliefs and desires can as well be referred to primitive organisms that only have a system of internal structures. However, the relevance to the explanation of the organism's behavior resides in what such structures indicate: they mean something and mean something "to" the organism of which they are part. In other words, Dretske thinks that intentionality is not unique of mental states, but quite ubiquitous in living and even non-living systems. For example, a thermometer refers to the temperature. Having content is then not unique to the human mind at all. Mental intentional states are actually somewhat limited compared to the intentional states of physical systems, as they miss a lot of information that physical systems would not miss. Paradoxically, the mind distorts the information that is available in the environment. Other systems are more faithful.

This argument can be summarized in terms of representations. The elements of a representational system have a content defined by what it is their function to indicate (what the British philosopher Paul Grice used to call "non-natural meaning"). Dretske distinguishes three types of representational systems: Type I have elements (symbols) that show no intrinsic power of representation (this includes maps, codes, etc); Type II have elements (signs) that are causally related to what they indicate (includes gauges); Type III (or natural) have their own intrinsic indicator functions (unlike Type I and Type II, in which humans are the source of the functions) and therefore a natural power of representation.

From these ideas Dretske developed a full-fledged theory of behavior. The term "behavior" is used in many different ways to mean different things. The behavior of an animal is commonly taken to be the actions it performs more or less by instinct or by nature. This is not necessarily "voluntary" behavior. The fact that women have menstruation is part of "female behavior", but it is not voluntary. Behavior is pervasive in nature, and cannot be restricted to animals: plants exhibit behavior too. Behavior is the production of some external effect by some internal cause. Behavior is a complex causal process wherein certain internal conditions produce certain external movements. First and foremost, behavior is a process. A process is caused by both a triggering cause (the reason why it occurs now) and a structural cause (the reason why the process is the way it is). This holds both for human behavior and the behavior of machines. For example, a thermostat turns on a furnace both because the temperature fell below a threshold and because it has been designed to turn on furnaces under certain conditions. In general, humans are interested in structural behavior, which in plants and animals has been determined by natural evolution and in machines has been built by humans.

In Dretske's view, intentionality is not a property useful to differentiate minds from matter, but a property that can help formalize the behavior of systems, human, biological and mechanical.

Phenomenology

In 1900 the German philosopher Edmund Husserl had expanded on Brentano's notion of intentionality. Since intentionality links mind and phenomena, he had concluded that phenomena and being are the same thing.

In the 1920s, the German philosopher Martin Heidegger, a follower of Husserl's "phenomenology", pointed out a fundamental flaw in dualist theories, and, consequently, in the whole mind-body debate.

In his opinion, Descartes' dualism is simply a consequence of a misleading vision of the world, according to which on one hand we have the "objective" world of "physical" reality (made of objects with physical properties) and on the other we have the "subjective" world of "mental" life (feelings, cognition, consciousness).

According to the Cartesian vision, the physical world is described by some objective facts that do not depend on our existence. We can perceive those facts and think about them. And we can act in the world based on our thoughts. But our relationship to the world and its objects is detached, observer-like.

Heidegger reminded us that, instead, we are part of that world. We are one of its "objects". We don't exist as independent entities, we exist as part of the world. There is no way that we can step back and, in a detached manner, watch what is happening: we are part of what is happening, and usually it is happening so fast that we don't even have time to think about it. We only have time to react by instinct.

Heidegger denies any value to the expressions "physical reality" and "mental life", and to the dichotomy objective/subjective: the world and the mind cannot be separated. Everything is subjective, or objective (depending on the definition), as everything that we know is our "interpretation" of what is happening in the world and we have no way of having an "objective" interpretation of that happening because we are part of it.

In our daily life we do not adopt a detached, logical approach to situations but we just "act". Usually, you analyze one of your actions only "after" you have performed it; and, typically, this happens only when something went wrong: you pause to reflect and analyze what and why something went wrong. Most of the time we are not "conscious" of why we are doing what we are doing.

Heidegger claimed that we are "thrown" into the world. Normally we don't "break down" the situation: we "break down" the world around us only when our actions fail and we need to find out why.

For example, we don't normally realize consciously what tools we are using to perform an action: a pair of scissors or a glass or a clutch stick. Only when our action fails, do we focus on the tool that we are using and why it is failing us.

When we are hammering a nail into wood, we are not interested in discussing the properties of the hammer and the nail and the wood: we just hammer. If it doesn't work, then we stop and analyze what is wrong with the hammer, the nail or the wood.

Same with the objects that surround us: we are rarely aware of every single object that surrounds us. But let's say that somebody locked us in a room and we needed to find a way out of there: only then would we "break down" the reality of that room into all of its objects, desperately looking for some help.

Sometimes when you suddenly focus on the drive to work, you get lost: all of a sudden, you don't recognize anymore the streets that you drive through every single morning. There are so many details that you never noticed: was there really a curve? is there really a billboard in that curve? And so forth. But if you don't focus on the route, you know perfectly well how to get to work.

If you close your eyes, you have to rediscover your own room, so many details of which are actually unconscious to you, even if you know it better than any other place in the world. When you try to move around your room blindfolded, you "break down" your knowledge about the room. The last time you had to do that was when you moved in.

In everyday life, we do not have a complete representation of the situation, and we cannot predict all the consequences of our actions; and we do not have time to search for either the representation or the prediction. Nonetheless we understand a situation and we act in it. And most of the times we survive. Only when our actions fail, do we need to step back and analyze the situation and try to figure out rationally why we failed. Logic is something that we use "after" the fact, to "troubleshoot" what we did wrong.

The science that we built to analyze the world is a complication. The truth is much simpler and so closer to our ordinary life.

According to Heidegger, there is a fundamental unity of the "Dasein" (of being). Subject and object cannot be separated. They cannot exist independently.

An individual is not a separate entity but a manifestation of Dasein in a world and within a tradition (society is a big component of that world).

We cannot study our beliefs as objects because we cannot abstract from them and observe them objectively. They are part of our belief system and every action we perform is affected by that same belief system, so we enter a vicious circle. We carry a burden of experience and knowledge with us which shapes our actions.

When we study something rationally, in a detached way, we are actually missing something by isolating it. Understanding something is being part of it. Cognition is praxis. We are "thrown" into the world and that's how we understand it and act in it. If we stop and observe it, then we may not be as good at acting in it.

Therefore Heidegger does not need mental representations to reason about. What matters is action: action of the world and action of us in the world. Representation is interpretation. There is no "objective" fact (or absolute truth) about the world.

Phenomenal Externalism

Fred Dretske ("Phenomenal externalism, or if meanings ain't in the head, where are qualia?", 1996) and the Canadian philosopher Ted Honderich expanded phenomenalism. The mind is not confined inside the skull, but it extends to include whatever tools are used. Not only is there no boundary between subject and object but, according to the US philosopher Andy Clark and the Australian philosopher David Chalmers ("The Extended Mind", 1998), the subject leaks out into the object: my mind literally spreads to the tools that i use in order to carry out my cognitive processes.

Variants on Materialism

The problem that has been haunting the endeavors of materialists for centuries is how the mental arises from the physical, how feelings originate from inanimate matter. A modern view is that the mind is indeed material, but somehow its material constituents behave differently from the matter that Physics has explained. Therefore, it is Physics that must be changed, or enlarged, to accommodate new types of natural phenomena.

John Searle's "biological naturalism" summarizes several of these materialistic opinions (despite being quite similar to property dualism). He thinks that (1) the mental is caused by neural processes and (2) the mental is a feature of the brain. We will understand consciousness when we know more about how the brain functions.

Brain processes cause mental states, but Searle objects to the common-sense conclusion that there must be both physical states and mental states, and therefore dualism. He views the mental state as a "feature of the brain". The mental state is an emerging property, not a separate substance. Hence no dualism; and no materialism.

Mental states are nonphysical, but form a novel class of features of the brain. Mental phenomena are irreducible to traditional Physics and Chemistry. Their properties (such as meaning and awareness) are different from those of matter.

The relation between brain states and mental states is causal, in both directions, each causing the other. Consciousness is an emergent property of the brain in the same way that properties of liquids emerge from those of the molecules they are made of. The mind is material, but at the same time it cannot be reduced to any other physical property.

Brains Cause Minds

The state of the "mind-body debate" can be appreciated by vivisecting some of its fundamental tenets.

Many contemporary philosophers, notably John Searle, would subscribe to the statement that minds are caused by brains. And the notion sounds intuitively true. A closer inspection reveals how unfounded this view is, and how misleading it can be for reasoning on consciousness. The problem is that the sentence is too informal to yield any formal, scientific discussion.

First of all, is the brain sufficient for a mind to exist? Can a brain alone yield a mind? We have no evidence whatsoever of a brain that, alone, causes a mind. Without a heart, would the brain cause a mind? Without the blood? Without the oxygen? Without all the nerves connecting to the senses? If somebody were to cut my head off and pull my brain out, I doubt that my brain would still cause my mind to exist. It would still be made of exactly the same matter, but it would not be able to cause a mind anymore. It would be, quite simply, rotting. The same object causes a mind or not depending on whether it is alive or dead. Truth is, we only have evidence of minds contained in bodies, and in living bodies. Therefore, it would be more appropriate to state that "living bodies cause minds".

Second, is the brain necessary for a mind to exist? We have no evidence of other (non-brain) things causing minds, but then we have no way of gathering that evidence. There is no way that we can know whether a different type of thing can also cause a mind. There is no way of knowing if an insect is conscious, if bacteria are conscious, if plants are conscious, if crystals are conscious, etc. Therefore, it would be more prudent to say that "at least living bodies cause minds". Which is a far less exciting proposition than "brains cause minds".

But the biggest problem is that even the term "brain" needs to be qualified. What is a brain? Is the skull part of the brain? Are the eyes part of the brain? What are the borders of the brain? Where does a nerve stop being part of the brain? Where do we cut off all the nerves, veins and muscles that link it to the rest of the body? At the chin? At the throat? How much can we change in a brain without changing its being a brain? What about other animals? Do cat's brains also qualify as brains? Do nervous systems of insects also qualify as brains? Does a computer qualify as a brain? Does a crystal qualify as a brain? What makes a brain a brain?

If minds are indeed caused by brains, what needs to be present in a conglomerate of neurons for it to become a brain that causes a mind? As a hypothetical Dr Frankenstein adds features to the ball of fibers that he assembled in his laboratory, at which point does that ugly mess become a mind that feels and thinks?

The closer we look, the less sure we are of the intuitive statement that brains cause minds.

The Cognition-consciousness Problem

The distinction between mind and body was clear in Descartes' times, but it is getting less obvious by the day as the physical and psychological sciences shed light on "mental" processes. Several of these processes are not exclusive of the mind (let alone of the human mind), but quite pervasive in nature. Remembering, learning, communicating are, to some extent, present in all forms of life. Since Descartes, the dilemma has been how do body and mind communicate. But, today, there is no mystery in how, say, learning communicates with the body: learning is a brain process that alters brain configurations in such a way that a different behavior will occur.

Today, we know that "body" extends to the brain, and brain is responsible for many phenomena that we consider mind and that are no more mysterious than the movement of a hand. Therefore, within the Cartesian dichotomy, "body" must be enlarged to encompass brain processes and "mind" must be restricted to conscious experience. Otherwise, most of the mystery is not a mystery at all: the way "mind" remembers or learns is no more mysterious than the way a muscle gets stronger or weaker. What is mysterious is that "remembering" and "learning" are sometimes associated with conscious experience. That is the real puzzle: how does a brain process of remembering (that is ultimately an electrochemical process of neurons triggering each other) communicate with our conscious life of feelings and emotions that seems to be located in a completely different dimension?

As David Chalmers pointed out, the paradox to be explained is not that body and mind communicate but that cognition and consciousness communicate.

Mind or Matter

It used to be a simple question: what is the soul? "Mind" complicated the question because it related the soul to a specific place, the brain, without being as specific. Is mind the soul? Is mind more than the soul? Is mind less than the soul?

The author of this book thinks that the problem is simply formulated in a nonscientific way. "Mind" is a generic term that refers to the set of cognitive faculties we humans have and sometimes (depending on the speaker) it also encompasses consciousness.

It would be more appropriate to focus on cognition itself. While some may be reluctant to credit animals with a mind, most will have no problem crediting them with some degree of cognitive faculties, such as memory, learning and even reasoning. Cognition can safely be assumed as a property of at least all living organisms, but a property that comes in (continuous) degrees: humans have "more" of it than, say, snails.

Furthermore, there are striking similarities between the behavior of cognitive (living) matter and the behavior of non-cognitive (inanimate) matter. Even a piece of paper exhibits a form of memory that resembles the way our memory works: if you bend it many times in the same direction, it will progressively "learn" to bend in that direction; if you stop bending it, it will slowly resume its flat position. Any piece of matter "remembers" what has happened to it in its shape, and sometimes in its chemical composition (that laboratory scientists can sometimes trace back in time). Far from being unique to the mind, cognitive faculties appear to be ubiquitous in nature.

Memory and learning can therefore be said to be ubiquitous in nature, as long as we assume that they come in degrees. Cognition may not necessarily be an exclusive property of living matter. Cognition may be a general property of matter, that the human brain simply amplifies to perform very interesting actions. At least that part of the mind, the one that has to do with cognitive faculties, may be "reduced" to material processes after all. The other part, consciousness, is a vastly more difficult topic.

The Factory of Illusions

"Thought" is an entirely different game. "Mind" defined as the totality of thoughts is a far more elusive mystery.

The mind is a factory of illusions. It creates an inner reality, as opposed to the outer reality of the world. We see colors and shapes, smell odors and perfumes, hear voices and sounds. We perceive the flowing of time. But the universe is made of particles and waves. The mind translates the world into sensations. Then it elaborates sensations to produce thoughts, memories, concepts, ideas. None of this is real. It is all one gigantic illusion. We will never even be sure whether anything exists at all.

Then the mind creates consciousness, i.e. the awareness of feeling those sensations and, among them, the subjective sensation of existing. May consciousness be the direct consequence of the existence of those illusions? Are all living beings equipped with sensory perception also endowed with consciousness?

Science needs crisp, reliable definitions, especially definitions of the objects it studies. Unfortunately, our mind is one of those things that we intuitively, "obviously" know, but, when we try to formalize, we realize we don't know at all. The most common way to define the mind is to list cognitive faculties: the mind is something that is capable of learning, remembering, reasoning, etc. The truth is that, by doing so, we have only shifted level: we now have to define learning, remembering, reasoning, etc. The more scientific we try to be, the more we end up with definitions that are broader than we would want them to be. As we saw, many things (and certainly many biological systems) can be said to be capable of some form of learning, remembering, reasoning, etc. Crystals exhibit sophisticated processes of self-organization.

What is so special about the mind? It is not the cognitive faculties. It is the inner life. The mind is a factory of illusions, that translates this world of particles and waves into a world of colors, sounds and smells. And it is the illusion of all illusions: consciousness. Therein lies the secret of the mind.

Further Reading

Armstrong, David Malet: A MATERIALIST THEORY OF THE MIND (Humanities Press, 1968)
Armstrong, David Malet: THE NATURE OF MIND (Cornell Univ Press, 1981)
Armstrong, David Malet: THE MIND-BODY PROBLEM (Westview, 1999)
Bechtel William: PHILOSOPHY OF MIND (Lawrence Erlbaum, 1988)
Block, Ned: READINGS IN PHILOSOPHY OF PSYCHOLOGY (Harvard Univ Press, 1980)
Bonnet, Charles: ESSAI DE PSYCHOLOGIE (1754)
Brentano, Franz: PSYCHOLOGY FROM AN EMPIRICAL STANDPOINT (1874)
Broad, Charlie Dunbar: THE MIND AND ITS PLACE IN NATURE (1925)
Campbell, Keith: Body and Mind (Macmillan, 1970)
Chalmers, David: THE CONSCIOUS MIND (Oxford University Press, 1996)
Chalmers, David: THE CHARACTER OF CONSCIOUSNESS (Oxford University Press, 2010)
Chomsky, Noam: LANGUAGE AND THOUGHT (Moyer Bell, 1991)
Churchland, Paul: MATTER AND CONSCIOUSNESS (MIT Press, 1984)
Clark, Andy: SUPERSIZING THE MIND (Oxford University Press, 2008)
Crane, Tim: THE MECHANICAL MIND (Penguin, 1995)
Davidson, Donald: INQUIRIES INTO TRUTH AND INTERPRETATION (Clarendon Press, 1984)
Dennett, Daniel: CONTENT AND CONSCIOUSNESS (Routledge, 1969)
Dennett, Daniel: BRAINSTORMS (Bradford, 1978)
Dennett, Daniel: KINDS OF MINDS (Basic, 1998)
Descartes, Rene': PRINCIPIA PHILOSOPHIAE (1644)
Dretske, Fred: KNOWLEDGE AND THE FLOW OF INFORMATION (MIT Press, 1981)
Dretske, Fred: EXPLAINING BEHAVIOR (MIT Press, 1988)
Dretske, Fred: NATURALIZING THE MIND (MIT Press, 1995)
Dretske, Fred: PERCEPTION, KNOWLEDGE AND BELIEF (Cambridge University Press, 2000)
Eccles, John: EVOLUTION OF THE BRAIN (Routledge, 1991)
Eccles, John & Popper, Karl: THE SELF AND ITS BRAIN (Springer, 1977)

Feigl, Herbert: THE MENTAL AND THE PHYSICAL (Univ of Minnesota Press, 1967)
Fodor, Jerry: LANGUAGE OF THOUGHT (Crowell, 1975)
Fodor, Jerry: REPRESENTATIONS (MIT Press, 1981)
Fodor, Jerry: MODULARITY OF THE MIND (MIT Press, 1983)
Fodor, Jerry: THE ELM AND THE EXPERT (MIT Press, 1994)
Gardner, Howard: MIND'S NEW SCIENCE (Basic, 1985)
Gregory, Richard: OXFORD COMPANION TO THE MIND (Oxford, 1987)
Heidegger, Martin: BEING AND TIME (1962)
Honderich, Ted: ON CONSCIOUSNESS (Edinburgh University Press, 2004)
Hume, David: A TREATISE OF HUMAN NATURE (1739)
Husserl Edmund: LOGICAL INVESTIGATIONS (1900)
Kim Jaegwon: SUPERVENIENCE AND MIND (Cambridge University Press, 1993)
Kim, Jaegwon: MIND IN A PHYSICAL WORLD (MIT Press, 1998)
Kripke, Saul: NAMING AND NECESSITY (Harvard Univ Press, 1980)
Leibniz, Gottfried: THE MONADOLOGY (1714)
Lewis, David K.: PHILOSOPHICAL PAPERS (Oxford Press, 1983)
Lewis, David K.: ON THE PLURALITY OF WORLDS (Basil Blackwell, 1986)
Lycan, William: CONSCIOUSNESS (MIT Press, 1987)
Lycan, William: MIND AND COGNITION (MIT Press, 1990)
Mach, Ernst: THE ANALYSIS OF SENSATIONS AND THE RELATION OF PHYSICAL TO THE PSYCHICAL (1886)
McGinn, Colin: CHARACTER OF MIND (Oxford Univ Press, 1997)
Minsky, Marvin: THE SOCIETY OF MIND (Simon & Schuster, 1988)
Popper, Karl & Eccles John: THE SELF AND ITS BRAIN (Springer-Verlag, 1977)
Popper, Karl: KNOWLEDGE AND THE BODY-MIND PROBLEM (Routledge, 1994)
Priest, Stephen: THEORIES OF THE MIND (Houghton Mifflin, 1991)
Putnam, Hilary: MIND, LANGUAGE AND REALITY (Cambridge Univ Press, 1975)
Quine, Willard: WORD AND OBJECT (1960)
Rosenthal, David: NATURE OF MIND (Oxford University Press, 1991)
Rucker, Rudy: INFINITY AND THE MIND (Birkhauser, 1982)
Russell, Bertrand: ANALYSIS OF MIND (1921)
Russell, Bertrand: ANALYSIS OF MATTER (Allen and Unwin, 1927)
Russell, Bertrand: AN INQUIRY INTO MEANING AND TRUTH (Penguin, 1962)

Ryle, Gilbert: THE CONCEPT OF MIND (Hutchinson, 1949)
Searle, John: THE REDISCOVERY OF THE MIND (MIT Press, 1992)
Sterelny, Kim: THE REPRESENTATIONAL THEORY OF MIND (Blackwell, 1990)
Stich, Stephen: FROM FOLK PSYCHOLOGY TO COGNITIVE SCIENCE (MIT Press, 1983)
Stich, Stephen: DECONSTRUCTING THE MIND (Oxford Univ Press, 1996)
Whitehead, Alfred: PROCESS AND REALITY (1929)
Wittgenstein, Ludwig: PHILOSOPHICAL INVESTIGATIONS (Macmillan, 1953)

COGNITION: A GENERAL PROPERTY OF MATTER

Cognition

Cognition is the set of faculties that allow the mind to process stimuli from the external world and to determine action in the external world. They comprise of perception, learning, memory, reasoning and so forth. Basically, we perceive something, we store it in memory, we retrieve related information, we process the whole, we learn something, we store it in memory, we use it to decide what to do next. All of these functions make up cognition.

Is all of cognition conscious? Is there something that we remember, learn or process without being aware of it? Probably. At least, the level of awareness may vary wildly. Sometimes we study a poem until we can remember all the words in the exact order: that requires a lot of awareness. Sometimes we simply store an event without paying too much attention to it. Consciousness is like another dimension. One can be engaged in this or that cognitive task (first dimension) and then it can be aware of it at different levels of intensity (second dimension). It is, therefore, likely that cognitive faculties and consciousness are independent processes.

Since it processes inputs and yields outputs, cognition has the invaluable advantage that it lends itself to modeling and testing endeavours, in a more scientific fashion than studies on consciousness.

Language too is a cognitive process. Given its importance for humans, it deserves a separate treatment, but it is likely that language's fundamental mechanisms are closely related to the mechanisms that support the other faculties.

Mediation

Over the last few decades, psychologists have been deeply influenced by the architecture of the computer. When it appeared, it was immediately apparent that the computer was capable of performing sophisticated tasks that went beyond mere Arithmetic, although they were performed by a complex layering of arithmetic sub-tasks. The fact that the computer architecture was able to achieve so much with so little led to the belief that the human mind could also be reduced to a rational architecture of interacting modules and sequential processes of computation.

In the second half of the 19th century, the German physiologist and physicist Hermann Helmholtz pioneered modern thinking about cognition when he advanced his theory that perception and action were mediated by a (relatively slow) process in the brain. The "reaction time" of a human being is high because neural conduction is slow. His studies emphasized that the stimulus must first be delivered to the brain and the idea of action must first be delivered to the limbs before anything can occur. Helmholtz thought that humans have no innate knowledge, that all our knowledge comes from experience. Perceptions are derived from unconscious inference on sense data: our senses send signals to the brain, which are interpreted by the brain and then turned by the brain into knowledge. Perceptions are mere hypotheses on the world, which may well be wrong (as proven by optical illusions). Perceptions are hypotheses based on our knowledge. Knowledge is acquired from perceptions. This paradigm became the "classical" paradigm of cognition.

Representation
The British psychologist Kenneth Craik speculated in 1943 that the human mind may be a particular type of machine which is capable of building internal models of the world (representing knowledge) and of processing them to produce action (making inferences). Craik's improvement over Descartes' automaton (limited to mechanical reactions to external stimuli) was considerable because it involved the ideas of an "internal representation" and a "symbolic processing" of such representation. Descartes' automaton had no need for knowledge and inference. Craik's automaton needs knowledge and inference. It is the inferential processing of knowledge that yields intelligence.

Symbol Processing
Craik's ideas predated the theory of knowledge-based systems, which was formalized when the US economist Herbert Simon and the US mathematician Alan Newell developed their theory of "physical symbol systems" ("Computer Science as Empirical Inquiry", 1976). Both the computer and the mind belong to the category of physical symbol systems, systems that process symbols to achieve a goal. A physical symbol system is quite simple: the complexity of its behavior is due to the complexity of the environment it has to cope with.

It was their belief that no complex system can survive unless it is organized as a hierarchy of subsystems. The entire universe must be hierarchical, otherwise it would not exist.

Production
Soon, the most abused model of cognitive psychology became one in which a memory system containing knowledge is operated upon by an inference engine; the results are added to the knowledge base and the cycle resumes indefinitely. For example, I may infer from my knowledge that it

is going to rain and therefore add to my knowledge base that there is a need for umbrellas.

In this fashion, knowledge is continuously created, and pieces of it represent solutions to problems. Every new piece of knowledge, whether acquired from the external world or inferred from the existing knowledge, may trigger any number of inferential processes, which can proceed in parallel. Since knowledge is mainly represented via "production rules" (rules that state that something becomes true when something else has become true), these systems are referred to as "production systems". A production rule is, ultimately, a formula of classical Logic.

John Anderson's ACT (1976) was a cognitive architecture capable of dealing with both declarative knowledge (represented by propositional networks) and procedural knowledge (represented by production rules). Declarative knowledge ("knowing that") can be consulted, whereas procedural knowledge ("knowing how") must be enacted in order to be used.

The relationship between the two types of knowledge is twofold. On one hand, the production system acts as the interpreter of the propositional network to determine action. On the other hand, knowledge is continuously compiled into more and more complex procedural chunks through an incremental process of transformation of declarative knowledge into procedural knowledge. Complex cognitive skills can develop from a simple architecture, as new production rules are continuously learned.

Anderson, therefore, thought of a cognitive system as having two short-term memories: a "declarative" memory (that remembers experience) and a "procedural" memory (that remembers rules learned from experience).

Anderson also developed a probabilistic method to explain how categories are built and how prototypes are chosen. Anderson's model maximizes the "inferential potential" of categories (i.e., their "usefulness"): the more a category helps predict the features of an object, the more the existence of that category makes sense. For each new object, Anderson's model computes the probability that the object belongs to one of the known categories and the probability that it belongs to a new category: if the latter is greater than the former, a new category is created.

Later editions of the architecture organize knowledge in three levels: a knowledge level (information acquired from the environment plus innate principles of inference), an algorithmic level (internal deductions, inductions and compilations) and an implementation level (setting parameters for the encoding of specific pieces of information).

Newell, working with John Laird and Paul Rosenbloom, proposed a similar architecture, SOAR (1987), based on three powerful concepts. The "universal weak method" is an organizational framework whereby knowledge determines the inferential methods employed to solve the problem, i.e. knowledge controls the behavior of the rational agent. "Universal sub-goaling" is a process whereby goals can be created automatically to deal with the difficulties that the rational agent encounters during problem solving. A model of practice is developed based on the concept of "chunking", the creation of new production rules, a process which is calibrated to produce the "power law of practice" that characterizes the improvements in human performance during practice at a given skill: the more you practice, the better you get at it. Within SOAR, each task has a goal hierarchy. When a goal is successfully completed, a chunk that represents the results of the task is created. In the next instance of that task, the system will not need to fully process it because the corresponding chunk already contains the instructions to achieve its goal. The process of chunking proceeds bottom-up in the goal hierarchy. The process will eventually lead to a chunk for the top-level goal for every situation that it can encounter.

These production systems are architectures advanced by the proponents of the symbolic processing approach in order to explain how the mind goes about acting, solving problems and learning how to solve new problems.

Mental Modules

Vision is one of the most important and complex cognitive faculties.

In 1982 the British psychologist David Marr delivered a landmark study on vision and, along the way, devised an influential cognitive architecture. Marr concluded that our vision system must employ innate information to decipher the ambiguous signals that it perceives from the world.

Processing of perceptual data must be performed by "modules", each specialized in some function, which are controlled by a central module. In a fashion similar to the linguist Noam Chomsky and the philosopher Jerry Fodor, Marr assumes that the brain must contain semantic representations that are innate and universal (i.e., of biological nature) in the form of modules that are automatically activated. The processing of such representations is purely syntactical. Marr, Chomsky and Fodor advanced the same theory of the mind, albeit from three different perspectives: they all believe that the mind can be decomposed in modules, and they all believe that syntactic processing can account for what the mind does.

Specifically, Marr explained the cognitive faculty of vision as a process in several steps. First, the physical stimulus from the world is received (in the form of physical energy) by transducers, that transform it into a symbol (in the form of a neural code) and pass it on to the input modules. Then these modules extract information and send it to the central module in charge of higher cognitive tasks.

Each module corresponds to neural subsystems in the brain. The central module exhibits the property of being "isotropic" (able to build hypotheses based on any available knowledge) and "Quinian" (the degree of confirmation assigned to a hypothesis is conditioned by the entire system of beliefs).

The visual system is thus decomposed in a number of independent subsystems. They provide a representation of the visual scene at three different levels of abstraction: the "primal sketch", which is a symbolic representation drawn from the meaningful features of the image (anything causing sudden discontinuities in light intensity, such as boundaries, contours, shading, textures); a two-and-a-half dimensional sketch, which is a representation centered on the visual system of the observer (e.g., describing the surrounding surfaces and their properties, mainly distances and orientation) and computed by a set of modules specialized in parameters of motion, shape, color, etc.; and finally the tri-dimensional representation, which is centered on the object and is computed according to some rules (Shimon Ullman's "correspondence rules").

This final representation is what is used for memory purposes. Not what the retina picked up, but what the brain computed.

Dimensions of Cognition

The obvious criticism against production systems is that they don't "look like" our brain.

David Marr claimed that a scientist can choose either of three levels of analysis: the computational level (which mathematical function the system must compute, i.e. an account of human competence), the algorithmic level (which algorithm must be used, i.e. an account of human performance) or the physical level (which mechanism must implement the algorithm). Different sciences correspond to different levels: Cognitive Science studies the mind at the computational level, Neurology studies the mind at the physical level, Eye Medicine studies the mind at the algorithmic level.

Newell refined that vision by dividing cognition into several levels. The program level represents and manipulates the world in the form of symbols. The knowledge level is built on top of the symbolic level and is the level of rational agents: an agent is defined by a body of knowledge, and has some goals to achieve and some actions that it can perform. An agent's behavior is determined by the "principle of rationality": the agent performs those actions that, on the basis of the knowledge it has, will bring it closer to the goals. General intelligent behavior requires symbol-level systems and knowledge-level systems.

Newell then broadened his division of cognitive levels by including physical and biological states.

The whole band can be divided into four bands: neural, cognitive, rational and social.

The cognitive band can be divided based on response times: at the memory level, the response time (the time required to retrieve the referent of a symbol) is about 10 milliseconds; at the decision level the response time is 100 milliseconds (the time required to manipulate knowledge), at the compositional level it is one second (time required to build actions), at the execution level it is 10 seconds (time required to perform the action).

At the rational band the system appears as a goal-driven organism, capable of processing knowledge and of exhibiting adaptive behavior.

Mental Models

The British psychologist Philip Johnson-Laird has questioned both the plausibility and the adequacy of a cognitive model based on production rules. A mind that only used production rules, i.e. Logic, would behave in a fundamentally different way from ours. People often make mistakes with deductive inference because it is not a natural way of thinking. The natural way is to construct mental models of the premises: a model of discourse has a structure that corresponds directly to the structure of the state of affairs that the discourse describes. For the same reason children are able to acquire inferential capabilities before they have any inferential notions: children solve problems by building mental models that are more and more complex, not by applying the rules of classical Logic (that they have not learned yet).

In his view, the mind represents and processes models of the world. The mind solves problems without any need to use logical reasoning. A sentence is a procedure to build, modify, extend a mental model. The mental model created by a discourse exhibits a structure that corresponds directly to the structure of the world described by the discourse. To perform an inference on a problem the mind needs to build the situation described by its premises. Such mental model simplifies reality and allows the mind to find an "adequate" (not necessarily "exact") solution.

Johnson-Laird's theory admits three types of representation: "propositions" (which represent the world through sequences of symbols), "mental models" (which are structurally analogous to the world) and "images" (which are perceptive correlates of models). Images are ways to approach models. They represent the perceivable features of the corresponding objects of the real world. Models, images and propositions are functionally and structurally different. Linguistic expressions are first transformed into propositional representations. The semantics of the mental language then creates correspondences between propositional representations and mental models, i.e. propositional representations are interpreted in mental models.

But the key to understanding how the mind works is in the mental models.

The French linguist Gilles Fauconnier advocates a similar vision in his theory of "mental spaces". Mental spaces proliferate as we think or talk. The mappings that link mental spaces, especially analogical mappings, play a central role in building our mental life. In particular, "conceptual blending" is a cognitive process which can be detected in many different cognitive, cultural and social activities. By merging different inputs, it creates a blended mental space that lends itself to what we call "creative" thinking. Therefore, Fauconnier finds that the same principles that operate at the level of meaning construction operate also at the level of scientific and artistic action.

The US linguist George Lakoff has given mental spaces an internal structure with his theory of "idealized cognitive models" that are embodied, i.e. they are linked with bodily experience.

Mental Imagery

"Mental imagery" is seeing something in the absence of any sensory signal, such as visualizing an object that is not actually present. The mystery is what is seen if in the brain there is no such image. When I stare at an object, i "see" the image that the visual system creates in the brain (whatever projection of dots through the retina to this or that region of the brain). But am i "seeing" when i am simply imagining a Ferrari?

Scientists have found no pictures or images in the brain, no internal eye to view pictures stored in memory and no means to manipulate them. Nevertheless, there is an obvious correspondence between a mental image of an object and the object.

The US psychologist Ronald Finke, for example, has identified five principles of equivalence between a mental image and the perceived object: the principle of implicit encoding (information about the properties of an object can be retrieved from its mental image), the principle of spatial equivalence (parts of a mental image are arranged in a way that corresponds to the way that the parts of the physical object are arranged), the principle of perceptual equivalence (similar processes are activated in the brain when the objects are imagined as when they are perceived), the principle of transformational equivalence (imagined transformations and physical transformations are governed by the same laws of motion), the principle of structural equivalence (the mental imagery exhibits structural features corresponding to those of the perceived object such that the relations between the object's parts can be both preserved and interpreted).

During the 1980s the debate became polarized around two main schools of thought: either (The US psychologist Stephen Kosslyn) the brain maintains mental pictures that somehow represent the real-world images, or (the US psychologist Zenon Pylyshyn) the brain represents images through a "non-imaginistic" system, namely language, i.e. all mental representations are descriptive and not pictorial (there are no picture-like representations in the brain).

Kosslyn put forth a representational theory of the mind of a "pictorial" type, as opposed to Jerry Fodor's propositional theory and related to Philip Johnson-Laird's mental models. Kosslyn thinks that the brain builds visual representations, which are coded in parts of the brain, and which reflect what they represent. Mental imagery involves scanning an internal picture-like entity. Mental images can be inspected and classified using pretty much the same processes used to inspect and classify visual perceptions. For example, they can be transformed (rotated, enlarged, reduced).

There exist two levels of visual representation: a "geometric" level, which allows one to mentally manipulate images, and an "algebraic" one, which allows one to "talk" about those images.

Kosslyn thinks that mental imagery achieves two goals: retrieve properties of objects, and predict what would happen if the body or the objects should move in a given way. Reasoning on shapes and dimensions is far faster when we employ mental images than concepts.

Kosslyn's is a theory of high-level vision in which perception and representation are inextricably linked. Visual perception (visual object identification) and visual mental imagery share common mechanisms.

Opposed to Kosslyn's "pictorialism" is Pylyshyn's "descriptionalism". Pylyshyn believes in a variant of Fodor's language of thought: to him images are simply the product of the manipulation of knowledge encoded in the form of propositions.

The "dual coding theory" of the Canadian psychologist Allan Paivio mediates these positions because it argues that the mind may use two different types of representation, a verbal one and a visual one, corresponding to the brain's two main perceptive systems. They both "encode" memories, but they do so in different ways (codes).

The neural processes that correspond to mental imagery in seeing people have also been detected in the brains of congenitally blind people. Thus mental imagery cannot possibly depend on forming a mental reconstruction of former visual sense experiences.

Mindsight

British philosopher Colin McGinn believes that percepts (the actual seeing of an object) and images (visualizing the object in the mind) are basically different "substances". McGinn comes up with a list of (nine) properties that differentiate them: percepts are unwanted (I see a tree when I see a tree, not when I want to see a tree), percepts contain information (images come from our minds and therefore we already have whatever information we put into visualizing them), percepts are located somewhere relatively to our body (whereas images are located in an abstract space of the mind), percepts don't come alone but with a background each point of which is in turn a percept, percepts exist whether we focus on them or not (whereas images exist only when we consciously visualize them), percepts are prone to error (I may recognize somebody who in fact was somebody else, whereas images are error-free because they are images of what I want to visualize), percepts do not block thinking about them (images do block thinking about them), etc.

He neglects one that most people would consider the most obvious one: I can perceive things that I have never imagined, but i cannot imagine things that i have never perceived. (I can construct images of objects that i have never seen, but those constructions are made out of objects that i have seen).

Furthermore he is not fair to images. I do visualize things and people without wanting to. If i stumble on to the name of a friend while i am reading a novel, i can't help visualizing my friend. It is not my "will" that visualizes my friend, but some kind of conditional reflex. Thus a percept may "force" an image, and the relationship between percept and image is not so obvious (in this case it is the relationship between a five-letter word and a human face).

McGinn thinks that images constitute the core of our cognitive life. Dreams, for examples, are complex systems of mental images. He has to posit a split in the self, a dream producer versus a dream consumer, because he has posited that images are active (dreams do not look active, they look more like percepts: the way we react to dreams is exactly the way we react to percepts). His idea is that dreams are passive for the self qua consumer.

McGinn also believes that the development of logic and language was triggered by the emergence of imagination: in order to understand a sentence one has to be capable of "imagining" its content. Imagination came first, meaning came later. This sounds more convincing, although he spends precious little time on such a monumental topic.

Imagination thus becomes a fundamental cognitive faculty, the one on which the most sophisticated cognitive faculties depend. The ability to create and manipulate mental images is, in a sense, what gives us an inner life.

One big problem with mental images is that we call them "images". In reality, when I recall a friend, I also recall his voice, and, when I recall an ice cream, I also recall its taste. I even recall my dislike for spiders whenever I recall a spider. Thus "mental images" are actually not images at all. They are indeed something very different from visual percepts because... they are not "visual".

Biological Epistemology

Solving a problem consists in visualizing it correctly. In his opinion the brain produces many different visualizations of the problem until one "fits". Then the solution is obvious. The brain produces a large number of ideas. Those that are "useless" are weakened; those that are useful are reinforced. The US psychologist Donald Campbell ("Blind variation and selective retention in creative thought as in other knowledge processes", 1960) viewed a selectionist process (blind variation and selective retention) at work in all the brain functions, from perception (recognizing that something is something) to problem solving.

At all levels the brain does not really "know" what to do: it just guesses, and the correct guesses are rewarded. Thinking originates from a population of guesses that evolve based on their usefulness or uselessness.

Creative thinking is a Darwinian process involving three processes: the generation of blind variations of ideas, the selection of the ideas that work best, and the reproduction of the selected ideas.

This is, basically, a generalization of Allen Newell's and Herbert Simon's "General Problem Solver" (1957),

Semantics and Pragmatics of Vision

A new paradigm for the study of human vision was introduced by the Canadian physiologist Melvyn Goodale and the British psychologist David Milner. They analyzed the visual pathways in the cerebral cortex (the "visual cortex") and realized that vision is actually a binary operation made of dual processes: on one hand is the conscious visual experience of the world, on the other hand is the visual control of unconscious (instinctive) action. They both require the eye as the organ, but they are functionally and structurally different processes.

Sensory information received from the eye diverges into two streams (two anatomically different pathways) when it leaves the visual cortex: a "ventral" stream flows from the primary visual cortex towards the inferior temporal lobe, while the "dorsal" stream flows from the primary visual cortex towards the posterior parietal lobe. This means that the ventral stream analyzes what object the eye is seeing, whereas the dorsal stream analyzes the spatial location of the object. In other words, the ventral stream is about recognition (e.g. of faces or objects) and conscious perception, whereas the dorsal stream is about automatic, unconscious action in space directed towards the object (typically action by the hand).

In a sense, there exist two kinds of vision: conscious perception and unconscious action. They are physically handled by two separate systems in the brain. There isn't a visual system: there are two visual systems that work parallel to each other.

Studying the difference between these two functional roles of the visual system, the French neurologist Marc Jeannerod had advanced the theory that there are two information processing systems for visual input: the "semantic processing system" (that yields a perceptual representation of the object) and the "pragmatic processing system" (that yields a motor representation of the same object).

The "dorsal" visual system that is common to most animals helps the animal carry out the orienting function: to detect movement (typically, either danger or food) and to guide action. The ventral visual system of the human brain is relatively distinct from the dorsal system and is related to the "semantic processing system" that analyzes and "recognizes" the object. This allows the human brain to carry out more sophisticated actions in response to a visual act. Human vision therefore originated when the functions of orientation and identification got separated.

The Frame

In the 1920s the German psychologist Otto Selz had a fundamental intuition: to solve a problem entails recognizing that the situation represented by the problem is described by a known "schema" and fill the gaps in the schema. A schema is a network of concepts that organize past experience.

Given a problem, the cognitive system searches the (long-term) memory for a schema that can represent it. Given the right schema, information in excess contains the solution.

Representation of past experience is a complete schema. Representation of present experience is a partially complete schema. By comparing the two representations (the complete schema that was created in the past with the partial schema that describes the current situation) one can infer (or, better, "anticipate") something relative to the present situation. For example, a schema tells us how to cross a street. Whenever we want to cross a street, we look for (i.e., we know that there must be) a traffic light. Thanks to the schema's anticipatory nature, to solve a problem is equivalent to comprehending it, and comprehending ultimately means reducing the current situation to a past situation.

In the 1960s Marvin Minsky rediscovered Selz's ideas: his "frame" is but a variation on Selz's schema.

A "frame" is a packet of information that helps recognize and understand a scene. It represents stereotypical situations and finds shortcuts to ordinary problems. A frame is the description of a category by means of a prototypical member plus a list of actions that can be performed on any member of the category. A prototype is described simply by a set of default properties. Default values, in practice, express a lack of information, which can be remedied by new information. Any other member of the category can be described by a similar frame that customizes some properties of the prototype.

Technically, a frame can provide multiple representations of an object: taxonomic (a conjunction of classification rules), descriptive (a conjunction of propositions of the default values) and functional (a proposition on the admissible predicates).

Memory is a network of frames, one for each known concept. Each perception selects a frame (i.e., classifies the current situation in a schema) which must then be adapted to that perception; and this is equivalent to interpreting the situation and deciding which action must be performed. Reasoning is adapting a frame to a situation. Knowledge imposes coherence on experience.

Because it does not separate cognitive phenomena such as perception, recognition, reasoning, understanding and memory which seem to occur always at the same time, the frame is more biologically plausible than other forms of knowledge representation that treat them as independent and sequential processes. Moreover, it offers computational advantages, because it focuses reasoning on the information that is relevant to the situation at hand.

Minsky later generalized the idea of the frame in a more ambitious model of how memory works. When a perception, or a problem-solving task, takes place, a data structure called "K-Line" (Knowledge Line) records the current activity (all the "agents" active at that time in memory). The recall of that event or problem is a process of rebuilding what was active (the agents that were active) in memory at that time. Agents are not all attached the same way to K-lines. Strong connections are made at a certain level of detail, the "level-band", whereas weaker connections are made at higher and lower levels. Weakly activated features correspond to assumptions by default, which stay active only as long as there are no conflicts. K-lines connect to K-lines and eventually form societies of their own.

The Script

In the 1970s, Roger Schank employed similar ideas in his model of "conceptual dependency" and in his theory of "case-based reasoning".

Case-based reasoning is a form of analogical reasoning in which the elementary unit is the "case", or situation. A type of memory called "episodic" archives generalizations of all known cases. Whenever a new case occurs, similar cases are retrieved from episodic memory. Then two things happen. First, the new case is interpreted based on any similar cases that were found in the episodic memory. Second, the new case is used, in turn, to further refine the generalizations, which are then stored again in episodic memory.

The crucial features of this model are similar to the ones that characterize frames. Interpretation of the new case is expectation-driven, based on what happened in previous cases. Episodic memory contains examples of solutions, rather than solutions or rules to find solutions.

Because the episodic memory is continuously refined, Schank refers to it more generally as "dynamic" memory: it can grow on its own, based on experience. The script is an extension of the idea of the case.

A scene is a general description of a setting and of a goal in that setting. A script is a particular instantiation of a scene (many scripts can be attached to one scene).

A script is a social variant of Minsky's frame. A script represents stereotypical knowledge of situations as a sequence of actions and a set of roles. Once the situation is recognized, the script prescribes the actions that are sensible and the roles that are likely to be played. The script helps understand the situation and predicts what will happen. A script performs "anticipatory" reasoning.

For example, a script describes the scene of a restaurant: the host seats the customers as they walk in and hands them a menu, the waiter takes their order and delivers it to the kitchen, the waiter brings the food, the waiter brings the bill, the waiter brings the change. When we enter a restaurant, we know what to expect. The moment we recognize a building as a restaurant, we know that there are waiters inside. If we walk in, we "expect" to be handed a menu, and we expect a waiter to take our order, and at the end we expect the bill. This is all in the script for restaurants. Our daily life is controlled by a multitude of scripts that direct our actions in all stereotypical situations (the vast majority of the situations that we encounter in our life).

A script is a generalization of a class of situations. If a situation falls into the context of a script, then an expectation is created by the script, based on what happened in all previous situations. If the expectation fails to materialize, then a new memory must be created. Such new memory is structured according to an "explanation" of the failure. Generalizations are created from two identical expectation failures. Memories are driven by expectation failures, by the attempt to explain each failure and learning from that experience. New experiences are stored only if they fail to conform to the expectations.

Here, again, remembering is closely related to understanding and learning. Memory has the passive function of remembering and the active function of predicting. The comprehension of the world and its categorization proceed together.

More and more complex structures were added by Schank and his associates to the basic model of scripts. A "memory organization packet" (MOP) is a structure that keeps information about how memories are linked in frequently occurring combinations. A MOP is both a storing structure and a processing structure. A MOP is basically an ordered set of scenes directed towards a goal. A MOP is more general than a script in that it can contain information about many settings (including many scripts). A "thematic organization packet" is an even higher-level structure that stores information independent of any setting.

Ultimately, knowledge (and intelligence itself) is stories. Cognitive skills emerge from discourse-related functions: conversation is "reminding" and storytelling is "understanding" (and in particular "generalizing"). The stories that are told differ from the stories that are in memory: in the process of being told, a story undergoes changes to reflect the intentions of the speaker. The mechanism underlying stories is similar to script-driven reasoning: understanding a story entails finding a story in memory that matches the new story and enhancing the old story with details from the new one. Underlying the mechanism is a process of "indexing" based on identifying five factors: theme, goal, plan, result and lesson. Memory actually contains only "gists" of stories, that can be turned into stories by a number of operations (distillation, combination, elaboration, creation, captioning, adaptation). Knowledge is embodied in stories and cognition is carried out in terms of stories that are already known.

This may explain both the passion for sport races (whether car racing or cycling) and for narrative art (whether films or novels). Both categories of human activities basically construct very complicated stories that challenge our minds. As we follow a race, we construct a story based on the stereotype actions that can happen in a race. And we root for our idol based on what stereotype actions would propel her/him to the head of the race. As we watch a film, we construct a story based on the stereotype actions that can happen in a film. And we identify with the protagonist based on what stereotype actions would make her/him succeed.

The Self-organizing Schema

Schemas resurface also in the work of the Australian mathematician Michael Arbib. Just like with Minsky's frames and Schank's scripts, Arbib argues that the mind constructs reality through a network of countless schemas. And, again, a schema is both a mental representation of the world and a process that determines action in the world.

Arbib's theory of schemas is based on two notions, one developed by the US mathematician Charles Peirce, and one due to the Swiss psychologist Jean Piaget. The first one is the notion of a "habit", a set of operational rules that, by exhibiting both stability and adaptability, lends itself to an evolutionary process. The second one is the notion of a schema, the generalizable characteristics of an action that allow the application of the same action to a different context (yet another variation on Selz). Both assume that schemas are compounded as they are built to yield successive levels of a cognitive hierarchy.

Arbib argues that categories are not innate, they are constructed through the individual's experience. What is innate is the process that underlies the construction of categories. Therefore, Arbib's view of the rules of categories is similar to Noam Chomsky's view of the rules of language.

Conceptual Graphs

Both frames and scripts are ultimately ways of representing concepts. A broader abstraction with a similar purpose has been proposed by the US mathematician John Sowa in his theory of "conceptual graphs", which is based both on Selz's schemas and on Peirce's "existential graphs" (his graph notation for logic).

A "conceptual graph" represents a memory structure generated by the process of perception. In practice, a conceptual graph describes the way percepts are assembled together. Conceptual relations describe the role that each percept plays.

Technically speaking, conceptual graphs are finite, connected, bipartite graphs (bipartite because they contain both concepts and conceptual relations, represented respectively by boxes and circles). Some concepts (concrete concepts) are associated with percepts for experiencing the world and with motor mechanisms for acting upon it. Some concepts are associated with the items of language. A concept has both a type and a referent. A hierarchy of concept-types defines the relationships between concepts at different levels of generality.

Formation rules ("copy", "restrict", "join" and "simplify") constitute a generative grammar for conceptual structures just like production rules constitute a generative grammar for syntactic structures. All deductions on conceptual graphs involve a combination of them.

The Forecasting Machine

Vernon Mountcastle, the US neurophysiologist who had discovered the columnar organization of the cerebral cortex, pointed out that the neocortex is uniform ("An Organizing Principle for Cerebral Function", 1978): there is no indication that any region of the neocortex operates differently from the other regions. That is puzzling because, at the end, some regions generate images and others generate sounds. The difference in "output", though, must be due almost entirely to the different inputs, because there is nothing inherently different between the regions that process one and the other. Furthermore, the inputs all comply with the exact same standard: they are all electrochemical signals. There is nothing inherently different in the format or in the processing. What is different is the pattern: spatial patterns, that are coincident patterns in time, versus temporal patterns. The vision of something and the hearing of something are created in the brain by analyzing a stream of patterns. It is the brain that sees, not the eye. It is the brain that hears, not the ear.

The US computer scientist Jeff Hawkins has revised Roger Schank's theory of scripts and its predictive model of memory. The brain doesn't simply process inputs and merge them at higher and higher levels, but at the same time does also the opposite: it predicts what comes next. This is how it can recognize events, situations and people, and understand what is going on. Memory is bidirectional: memory interprets the present while it is absorbing it. At the same time the memory of the past is tweaked every time we use it, based on the result of using it. Memory becomes a reservoir of possible solutions to problems. Memory constantly predicts what will happen next. That's how it can keep track of situations that would require almost infinite computational capabilities.

Hawkins believes that the columnar structure of the neocortex yields a hierarchical structure of mind (lower levels being closer to raw perception, and higher levels being closer to pure abstraction). At each level the basic form of processing is a combination of matching and creating patterns. Perception, action and cognition are spread throughout this hierarchy of patterns.

The brain's function is to store memories as patterns and to make predictions based on such patterns. The brain is fundamentally a forecasting machine, that continuously analyzes the present (current patterns) against the past (stored patterns) in order to predict the future.

Habits

Charles Peirce, the founder of pragmatism, once proposed a unifying view of matter and mind, although it was disguised as a theory of "habits".

Peirce believed that there was no absolute definition of things, including truth itself. An object is defined by the effects of its use: a definition that works well is a good definition. An object "is" its behavior. The meaning of a concept consists in its practical effects on our daily lives: if two ideas have the same practical effects on us, they have the same meaning. The meaning of a concept is a function of the relations among many concepts: a concept refers to an object only through the mediation of other concepts.

Truth is usefulness and validity: something is true if it can be used and validated. In practice, truth is defined by consensus of the society. Truth is not agreement with reality, it is agreement among humans (reached after a process of scientific investigation). Truth is "true enough", not necessarily an absolute, unchanging truth. Truth is a process, a process of self-verification.

What is relevant is not the concept of "true", but the "belief". We use beliefs in our daily lives, not theorems that prove what is true and what is false. Beliefs become fixed over a lifetime through experience and verification. I believe something if that belief has proven useful over the course of my life. Beliefs lead to the formation of habits that, in turn, get reinforced through experience. The more useful that belief turns out to be, the stronger it becomes in my mind. Ditto for habits: the better they work for me, the more "habitual" they become.

Peirce noted that the process of habit creation is pervasive in nature. All matter acquires habits. Matter is mind whose "beliefs" have been fixed to the extent that they can't be changed anymore. Habit is what makes objects what they are. An object is defined by the set of all its possible behaviors, i.e. by its "habits". I am my habits. It makes no sense to talk of something or someone who does not have habits: randomness is absence of an identity.

The laws of Physics describe the habits of matter, because what physicists observe is the habits of nature. For example, heavenly bodies have the habit of attracting each other, thus the laws of gravitation.

Systems evolve because of chance, which is inherent to the universe ("tychism"). Habits progressively remove chance from the universe. The universe is evolving from absolute chaos (chance and no habits) towards absolute order (all habits are fixed). One can see Darwinian evolution at work on systems towards stronger and stronger habits. Human beliefs are a particular case of habits, that also get fixed through experience.

Enter the Body

Largely inspired by French philosopher Maurice Merleau-Ponty, a number of US philosophers, such as Mark Johnson and Maxine Sheets-Johnstone, have "rediscovered" the body.

Cognitive scientists tend to focus on the input (perception) and neglect the output (action). Most models of the mind have been built by concentrating on perception: how the world is perceived by the mind. Very little is usually said on how the mind acts on the world. But action (movement) is no less important a part of our experience.

Sheets-Johnstone retraces Merleau-Ponty's philosophy in claiming that thinking is modeled on the body and grounded in animate form. The "tactile-kinesthetic body" is the source of corporeal concepts. All our cognitive life is grounded in movement. Consciousness does not arise from matter, but from self-movement. Even the simplest forms of life enjoy a "meta-corporeal consciousness of the chemical constitution of the environment".

The Process Behind the Structure

Robert Cairn once used an analogy: evolution is to biology what development is to psychology, i.e. the process behind the structure.

The US psychologists Esther Thelen and Linda Smith advanced a theory of development that was as opportunistic as evolution. Their emphasis is on processes of change, on the ever-active self-organizing processes of living systems that are analogous to the selection algorithms of evolution.

For them development is the outcome of the interplay between action and perception within a system that, by its thermodynamic nature, seeks stability. Performance and cognition emerge from this process of interaction between a system and its environment. Cognition, in particular, is an emergent structure, situated and embodied, just like any other skill of the organism.

The development of the brain seems to be orderly, incremental, and directional (towards nutritional independence and reproductive maturity), but this is an illusion. In reality development is not driven by a grand design: it is driven by opportunistic, syncretic and exploratory processes. At a closer look, in fact, development is modular and heterochronic (i.e., different organs develop at different rates and different times), although the organism progresses as a whole. Global regularities (and simplicity) somehow arise from local variabilities (and complexities).

Knowledge for thought and action (i.e., categories) emerges from the dynamics of pattern formation in the context of neural group selection. Perception, action and cognition are rooted in the same pattern formation processes. Categories arise (self-organize) spontaneously and reflect the experiences of acting and perceiving, i.e. of interacting with the world. More precisely, categories are created through the cross-relation of multimodal (hearing, seeing, feeling, etc) experiences. Unity of perception and action is reflected in the way categories are formed. Development can then be viewed as the dynamic selection of categories. Categories are the foundation of cognitive development.

At the same time, categories are but a specific case of pattern formation. Therefore, cognitive development is a direct consequence of the properties of nonlinear dynamic systems, i.e. of self-organizing complex systems.

Being in the world "selects" categories. Therefore meaning itself is emergent.

These features are shared by all organisms: every living system is a cognitive system.

The Unity of Cognition

Perception, memory, learning, reasoning, understanding and action are simply different aspects of the same process. This is the opinion implicitly stated by all schema-based models of cognition. All mental faculties are simply different descriptions of the same process, different ways of talking about the same thing: one, whole process of cognition. There is never perception without memory, never memory without learning, never learning without reasoning, never reasoning without understanding, and so forth. One happens because all happen at the same time.

The mind contains this powerful algorithm that operates on cognitive structures. That algorithm has been refined by natural selection to be capable of responding in optimal time. This might be the case partly because that algorithm operates on structures that already reflect the nature of our experience. Our experience occurs in situations, each situation being a complex aggregate of factors. The actions that we perform in a given situation are rather stereotyped. The main processing of the algorithm goes into recognizing the situation. Once the situation is recognized, somehow it is reduced to past experience and that helps figure out quite rapidly the appropriate action.

Needless to say, various levels of cognition can be identified in other animals, and even in plants. Even in crystals and rocks. Everything in nature can be said to remember and to learn, everything can be said to be about something else.

Cognition is not "all" there is in the mind: this is the utilitarian, pragmatic, mechanical part of the mind. The mind also has awareness. But consciousness does not seem to contribute much to the algorithm, does not seem to significantly affect the structure of past experience, does not seem to have much to do with our ability to deal with situations. A being with no consciousness, but with the same cognitive algorithm and the same cognitive structures (i.e., with the same cognitive architecture), would probably behave pretty much like us in pretty much all of our daily actions, without the emotions.

Cognition does not seem to require consciousness. Ultimately, it is simply a material process of self-organization. It seems possible to simulate this process by an algorithm, which means that cognition is not exclusive to conscious beings. It may well be possible to build machines that are cognitive systems. Cognition may actually turn out to be a general property of matter, of all matter, living and nonliving.

Further Reading
Anderson, John-Robert: THE ARCHITECTURE OF COGNITION (Harvard Univ Press, 1983)
Anderson, John-Robert: THE ADAPTIVE CHARACTER OF THOUGHT (Lawrence Erlbaum, 1990)
Anderson, John-Robert: RULES OF THE MIND (Lawrence Erlbaum, 1993)
Arbib, Michael & Hesse, Mary: CONSTRUCTION OF REALITY (Cambridge University Press, 1986)
Ballard, Dana: COMPUTER VISION (Prentice Hall, 1982)
Block, Ned: IMAGERY (MIT Press, 1981)
Craik, Kenneth: THE NATURE OF EXPLANATION (Cambridge Univ Press, 1943)
Fauconnier, Gilles: MENTAL SPACES (MIT Press, 1994)
Finke, Ronald: PRINCIPLES OF MENTAL IMAGERY (MIT Press, 1989)
Finke, Ronald: CREATIVE COGNITION (MIT Press, 1992)
Franklin, Stan: ARTIFICIAL MINDS (MIT Press, 1995)
Green, David: COGNITIVE SCIENCE (Blackwell, 1996)
Hampson, Peter & Morris, Peter: UNDERSTANDING COGNITION (Blackwell, 1995)
Hawkins, Jeff: ON INTELLIGENCE (2004)
Johnson-Laird, Philip: MENTAL MODELS (Harvard Univ Press, 1983)
Johnson-Laird, Philip: THE COMPUTER AND THE MIND (Harvard Univ Press, 1988)
Johnson-Laird, Philip & Byrne Ruth: DEDUCTION (Lawrence Erlbaum, 1991)
Johnson, Mark: THE BODY IN THE MIND (University of Chicago Press, 1987)
Kosslyn, Stephen: IMAGE AND MIND (Harvard University Press, 1980)
Kosslyn, Stephen: GHOSTS IN THE MIND'S MACHINE (W. Norton, 1983)
Kosslyn, Stephen & Koenig Olivier: WET MIND (Free Press, 1992)
Kosslyn, Stephen: IMAGE AND BRAIN (MIT Press, 1994)

Laird, John; Rosenbloom, Paul & Newell, Alan: UNIVERSAL SUBGOALING AND CHUNKING (Kluwer Academics, 1986)

Leyton ,Michael: SYMMETRY, CAUSALITY, MIND (MIT Press, 1992)

Luger, George: COGNITIVE SCIENCE (Academic Press, 1993)

Marr, David: VISION (MIT Press, 1982)

McGinn, Colin: MINDSIGHT (2004)

Goodale, Melvyn & Milner, David: THE VISUAL BRAIN IN ACTION (Oxford University Press, 1995)

Minsky, Marvin: SEMANTIC INFORMATION PROCESSING (MIT Press, 1968)

Minsky, Marvin: THE SOCIETY OF MIND (Simon & Schuster, 1985)

Newell, Allen: UNIFIED THEORIES OF COGNITION (Harvard Univ Press, 1990)

Paivio, Allan: IMAGERY AND VERBAL PROCESSES (Holt, Rinehart and Winston, 1971)

Posner, Michael: FOUNDATIONS OF COGNITIVE SCIENCE (MIT Press, 1989)

Pylyshyn, Zenon: COMPUTATION AND COGNITION (MIT Press, 1984)

Pylyshyn, Zenon: SEEING AND VISUALIZING (MIT Press, 2003)

Schank Roger: SCRIPTS, PLANS, GOALS, AND UNDERSTANDING (Lawrence Erlbaum, 1977)

Schank, Roger: DYNAMIC MEMORY (Cambridge Univ Press, 1982)

Selz, Otto: ON THE LAWS OF THE ORDERLY THOUGHT PROCESS (1913)

Sheets-Johnstone, Maxine: THE PRIMACY OF MOVEMENT (John Benjamins, 1981)

Sowa, John: CONCEPTUAL STRUCTURES (Addison-Wesley, 1984)

Stillings, Neil: COGNITIVE SCIENCE (MIT Press, 1995)

Thelen, Esther & Smith, Linda: A DYNAMIC SYSTEMS APPROACH TO THE DEVELOPMENT OF COGNITION AND ACTION (MIT Press, 1994)

Ullman, Shimon: THE INTERPRETATION OF VISUAL MOTION (MIT Press, 1979)

INSIDE THE BRAIN

Life is not the only field in which traditional Physics seems to be powerless to offer comprehensive explanations. The brain is another system that seems to obey laws that only partially reflect the linear universe implied by Physics. Explaining the evolution of life required a new paradigm, the paradigm of "design without a designer". It turns out that the functioning of the brain (that is responsible for the evolution of our thoughts) requires a similar paradigm.

A word of caution: everything we think about the brain comes from our brain. When I say something about the brain, it is my brain talking about itself.

Connectionism

Human memory may be deficient in many ways (it forgets, it does not remember "photographically"), but somehow it is extremely good at recognizing.

I'd recognize a friend even if he grew a beard, even if he's wearing different clothes every day, even if I see him sideways, and from any possible angle. How can I recognize all those images as the same image if they are all different? It is almost impossible to take the identical shot of a person twice: some details will always be different: how can I recognize that it is the same person, if the image is always different? I can show you two pictures of a street, taken at different times: you will recognize them as pictures of the same street. But there are probably countless differences: cars that were parked moved away and new cars took their places, pedestrians that were walking are gone, dogs and birds have changed positions, smoke has blown away, all the leaves of all the trees have moved because of the breeze, etc. How do you recognize that it is the same street, if the image of that street is never the same? Even a baby recognizes that an object turned sideways is still the same object, even if it looks completely different. Take a box and rotate it 45 degrees: it now looks like a completely different geometric figure. Nonetheless, a baby can recognize that it is the same object.

The key to understanding our memory may lie in the peculiar structure of our brain. Unlike most of our artifacts, which are designed to be modular, hierarchical and linear, a brain is an amazingly intricate piece of work. The brain does not work the way our artifacts work. There seems to be no "designer" that specifies what has to be designed. There seems to be a huge number of connected units, none of which prevails and all of which cooperate in some fashion to produce what transpires as "intelligent" behavior.

At the end of the 19th century, the US psychologist William James had a number of powerful intuitions: 1. That the brain is built to ensure survival in the world; 2. That cognitive faculties cannot be abstracted from the environment that they deal with; 3. That the brain is organized as an associative network; 4. That associations are governed by a rule of reinforcement. The latter two (3 and 4) laid the foundations for the "connectionist" model of the brain. The former two (1 and 2) laid the foundations for a cognitive model grounded in a Darwinian scenario of survival of the fittest.

Selective Behavior

Other psychologists contributed, directly or indirectly, to the connectionist model of the brain. In the 1920s behaviorist scientists such as the Russian physiologist Ivan Pavlov and the US psychologist Burrhus Skinner were influential in emphasizing the simple but pervasive law of learning through "conditioning": if an unconditioned stimulus (e.g., a bowl of meat) that normally causes an unconditioned response (e.g., the dog salivates) is repeatedly associated with a conditioned stimulus (e.g., a bell), the conditioned stimulus (the bell) will eventually cause the unconditioned response (the dog salivates) without any need for the unconditioned stimulus (the bowl of meat).

Behaviorists came to believe that all forms of learning could be reduced to conditioning phenomena. To Skinner, all learned behavior is the result of selective "reinforcement" of random responses. Mental states (what goes on in our minds) have no effect on our actions. Skinner did not deny the existence of mental states, he simply denied that they explain behavior. A person does what she does because she has been "conditioned" to do that, not because her mind decided so. Skinner noticed a similarity between reinforcement and natural selection: random mutations are "selected" by the environment, and random behavior is also selected by the environment. A random action can bring reward (from the environment) that will cause reinforcement and therefore will increase the chances that the action is repeated in the future. An action that does not bring reward will not be repeated.

The environment determines which behavior is learned, just like the environment determines which species evolve.

Cognition

Gestalt psychologists, instead, focused on higher cognitive processes and opposed the idea that the individual stimulus could cause an individual response. For example, in 1938 the German psychologist Max Wertheimer claimed that perception ought to be more than the sum of the perceived things, i.e. that the whole is more than the sum of the parts. He showed, for example, how one can alter the parts of a melody but the listener would still recognize the melody.

Perception of the whole does not depend on perception of all of its parts. We recognize the shape of a landscape way before we recognize each tree and rock in the landscape, and we recognize that a tree is a tree before we recognize what kind of tree it is, because recognizing the species requires an analysis of its parts.

Already in the 1920s the German psychologist Wolfgang Koehler had claimed that most problem-solving is not due to a decomposition of the problem but to sudden insight. Problems can be solved by visualizing the problem correctly so that "insight" shows the solution as obvious. One may not recognize a familiar face for a few seconds, and then suddenly recognize it. This is not due to a myriad calculations, but to a sudden insight that cannot be broken down into atomic processes. It is just a sudden insight.

The German neurologist Kurt Goldstein viewed the organism as a system that has to struggle in order to cope with the challenges of the environment and of its own body. The organism cannot be divided into "organs" and far less into "mind" and "body", because it is the whole that reacts to the environment. Nothing is independent within the organism. The organism is a whole.

"Disease" is a manifestation of a change of state between the organism and its environment. Healing does not come through "repair" but through adaptation. The organism cannot simply return to the state preceding the event that changed it, but has to adapt to the conditions that caused the new state. In particular, a local symptom is not meaningful to understand a "disease", and the organism's behavior during a disease is hardly explained as a response to that specific symptom. A patient's body will often undergo mass-scale adjustments. Goldstein emphasizes the ability of organisms to adjust to catastrophic breakdowns of their most vital (mental or physical) functions. The organism's reaction is often a redistribution of its (mental or physical) faculties.

Coherently, gestalt psychologists claimed that form is the elementary unit of perception. We do not construct a perception by analyzing a myriad data. We perceive the form as a whole.

Around 1950 experiments by the US neurologist Karl Lashley confirmed that intuition: a lesion in the brain does not necessarily cause a change in the response. Lashley concluded that functions are not localized but distributed around the brain, that there are no specialized areas, that all cortical areas are equally "potent" in carrying out mental functions (this was his "principle of equipotentiality"). Lashley realized that this architecture yields a tremendous advantage: the brain as a whole is "fault tolerant", because no single part is essential to the functioning of the whole.

Lashley also enunciated a second principle that can be viewed as its dual, the principle of "mass action": every brain region partakes (to some extent) in all brain processes. Lashley even imagined that memory behaved like an electromagnetic field and that a specific memory was a wave within that field. While he never came to appreciate the importance of the "connections", Lashley's ideas were sort of complementary to the ideas of connectionism.

Functions are indeed localized in the brain, but the processing of information inside the brain involves "mass action". The function of analyzing data from the retina is localized in a specific region of the brain, but the function of "seeing" is not localized, because it requires processes that are spread around the brain.

There are maps of the retina in the brain (even more than one), and there are maps of the entire body in the brain, and they are orderly maps. The brain keeps a map of what is going on in every part of the body.

The Primacy Of Connections

The US psychologist Edward Thorndike, a student of William James had already explained how Skinner's reinforcement occurs. Thorndike had been the first psychologist to propose that animals learn based on the outcome of their actions (the "law of effect") and Skinner simply generalized his ideas.

The "law of effect" stemmed from the observation that animals do not learn from the success of other animals; nor do they learn if guided to the correct solution. They learn based on how their own actions succeed or fail.

The "law of effect" states that the probability that a stimulus will cause a given response is proportional to the satisfaction that the response has produced in the past, and inversely proportional to the dissatisfaction. This principle was consistent with both natural selection and behaviorist conditioning/reinforcement. It complied with Darwinism because it assumed that responses were initially random and later "selected" by success or failure. It improved Pavlovian conditioning because it explained how new patterns of behavior could emerge.

Thorndike modeled the mind as a network of connections among its components. Learning occurs when elements are connected. A habit is nothing more than a chain of "stimulus-response" pairs. Behavior is due to the association of stimuli with responses that is generated through those connections. This model was also consistent with gestalt holism because in a vast network of connections the relative importance of an individual connection is negligible.

Connectionism can be viewed at various levels of the organization of the mind. At the lowest level, it deals with the neural structure of the brain. The brain is reduced to a network of interacting neurons. Each neuron is a fairly simple structure, whose main function is simply to transmit impulses to other neurons. When anything happens to a neuron, it is likely to affect thousands of other neurons because its effects can propagate very quickly from one neuron to the other.

From the outside, the only thing that matters is the response of the brain to a certain stimulus. But that response is the result of thousands of messages transmitted from neuron to neuron according to the available connections. A given response to a given stimulus occurs because the connections propagate that stimulus from the first layer of neurons to the rest of the connected neurons until eventually the response is generated by the last layers of neurons. As long as the connections are stable, a given stimulus will always generate the same response. When a connection changes, a different response may be produced. Connections change, in particular, when the brain "learns" something new. The brain "learns" what response is more appropriate to a given stimulus by adjusting the connections so that next time the stimulus will produce the desired response.

The functioning of the brain can be summarized as a continuous refining of the connections between neurons. Each connection can be strengthened or weakened by the messages that travel through it. In 1949 the Canadian physiologist Donald Hebb realized that strengthening and weakening of connections depend on how often they are used. If a connection is never used, it is likely to decay, just like any muscle that is not exercised. If it is used very often, it is likely to get reinforced. A Darwinian concept came to play a key role in the organization of the brain: competitive behavior. Connections "compete" to survive.

At a higher level, a connectionist organization can be found in the way our mind organizes concepts. Concepts are not independent of each other: a concept is very much defined by the other concepts it relates to. The best definition of a concept is probably in terms of other concepts and the way it relates to them. Concepts also rely on an associative network. Therefore, William James' four maxims also apply to concepts.

Ultimately, the connectionist model explained the functioning of the brain by employing the same paradigm that Darwin had used to explain the evolution of life: design without a designer.

The Neural Structure of the Brain

The human brain is probably the single most complex structure that we have found in the universe. Even the human genome is simpler.

First of all, the brain is really just the upper extremity of the spinal cord. Nerves departing from the spinal cord communicate with the rest of the body. The spinal cord contains the same gray matter of the brain.

Most of the human brain (the cerebrum) is made of two hemispheres, linked by the "corpus callosum", and covered by the cortex.

Under the corpus callosum is located one of the main areas of control of behavior, containing the "thalamus", the "hypothalamus" and the "amygdala". The thalamus is a mini-mirror of the cortex: it seems to replicate the same information, but on a smaller scale. The two amygdalae are widely believed to be in charge of emotions: affection, fear and attention originate or are amplified here. The function of the two thalami seems to be to convey signals from the senses to the cortex and from the cortex to the muscles. The amygdala has the power to take over this strategic highway.

The hypothalamus, located below the thalamus, is involved in many "autonomic" functions (heartbeat and breathing, but also hunger, lust, fear). It seems, in particular, to be responsible for controlling body temperature (pretty much like a thermostat). When warned by the immune system (via chemicals in the blood stream), that the body is being attacked, the hypothalamus triggers a simultaneous increase in body metabolism and a reduction in blood flow, i.e. a "fever".

The cortex is one of the main areas of sensory-motor control. The cortex is by far the largest structure in the brain: in humans, it accounts for about two thirds of the total brain mass. The terms "cortex" and "neocortex" are often used interchangeably because the neocortex constitutes most of the cerebral cortex in humans, but this is not true for other animals.

Located at the base of each hemisphere are the hippocampi. The hippocampus is one of the main areas for recalling long-term memory. It takes about three years to consolidate short-term memory into long term memory. For three years the hippocampus is directly responsible for retrieving a memory. After that period, the memory slides into long term memory. Lesions to the hippocampus result in forgetting everything that happened over the last three years and not being able to remember anything ever again for longer than a few seconds.

Alternatively, one can view a brain hemisphere as two concentric spheres: the inner one is the limbic system, comprising amygdala, thalamus, hypothalamus and hippocampus; the outer one is the neocortex. The neocortex processes sensory information and channels it to the hippocampus, which then communicates with the other organs of the limbic system. The limbic system appears to be a central processing unit that mediates between sensory input and motor output, between bodily sensations and body movements. In other words, the limbic system appears to be the main connection between mind and body. The limbic system is (evolutionarily speaking) the oldest part of the brain, the part that humans share with all mammals and that is well developed also in other vertebrates.

Each hemisphere's cerebral cortex can also be described as made of four lobes: the frontal lobe, that contains the primary motor area; the temporal lobe, that includes the hippocampus and is related to memory; the occipital lobe, concerned with vision; and the parietal lobe, important for spatial relationships and bodily sensations. The Indian neurologist Vilayanur Ramachandran picks three regions in the human brain that are many times larger than the corresponding areas in the chimp's brain: Wernicke's area in the left temporal lobe (where a lesion impairs the comprehension of language); the prefrontal cortex of the frontal lobe (where a lesion can totally alter the personality and obliterate any ethical sense); the right and left parietal lobe (responsible for spatial mapping and logical thinking).

Behind the hemispheres is the "cerebellum", one of the main areas of integration of stimuli and coordination of action. The cerebellum contains areas like the "pons" that communicate with the rest of the body. The cerebellum is a bit like a miniature brain: it is divided into hemispheres and has a cortex that surrounds these hemispheres. The cerebellum seems to be indispensable for coordinating complex actions such as playing a musical instrument.

Finally, the brainstem is the general term for the area of the brain between the thalamus and spinal cord. This is at the bottom of the brain, next to the cerebellum, and represents the brain's connection with the "autonomic" nervous system, the part of the nervous system that regulates functions such as heartbeat, breathing, sleeping, etc. These are mechanical functions, but even more vital than the higher ones.

Dominance
Since the "split-brain" studies carried out by the US psychologist Roger Sperry and Michael Gazzaniga ("Some functional effects of sectioning the cerebral commissures in man", 1962), it has been held that the two hemispheres of the human brain control different aspects of mental life

(Konstantin Bykov had already proved this in Russia in 1924): the left hemisphere is dominant for language and speech, the right brain excels at visual and motor tasks and may also be the prevalent source of emotions. This is due to the fact that the two hemispheres are not identical. For example, the speech area of the cortex is much larger in the left hemisphere. The roles of two hemispheres are not so rigid, though: a child whose left hemisphere is damaged will still learn to speak and will simply use the right hemisphere for language functions.

Just like it dominates in language, the left hemisphere also dominates in movement. Both hemispheres organize movement of limbs (each hemisphere takes care of the limbs at the opposite side of the body), but the left hemisphere is the one that directs the movement and that stores the feedback (the one that learns skills/ habits). If the two hemispheres are separated, the right limbs keep working normally, but the left limbs become clumsy and are often unable to carry out even simple learned skills like grabbing a glass.

Brain asymmetry is not uncommon in other species, but handedness (that individuals always prefer one hand over the other) is uniquely human, and handedness appears to depend on the asymmetry of the hemispheres.

The main "bridge" between the two hemispheres is the corpus callosum, but a number of other "commissures" (communication channels) exist, and their purpose is not known.

The US psychologist Ross Buck proposed a decomposition of human behavior that mirrors the "behavior" of each hemisphere. In his view, our behavior is the product of several systems of organization which belong to two big families. The first one is the family of innate special-purpose processing systems (reflexes, instincts, etc.). In general their function is "bodily adaptation" to the environment. In general, their approach is not analytic but holistic and syncretic: they don't "reduce" the situation to its details, they treat it as a whole. In Buck's view, these processes are innate, we don't need to learn them. The second family contains acquired general-purpose processing systems. In general their function is to make sense of the environment. Their approach is sequential and analytic. The former family is associated with the right hemisphere of the brain and is responsible for emotional expression; the latter is associated with the left hemisphere and is responsible for symbolic thinking. The two families cooperate in determining the body's behavior.

Asymmetry

The central nervous system of all organisms is remarkably symmetrical. That symmetry ends somewhere inside the mammalian brain, as lesions to the same location in different brain hemispheres have proven, particularly the neurological disorder of aphasia. (The loss of language skills, always related to damage to the left hemisphere, a correlation first studied by the German neurologist Karl Wernicke in 1874, and the loss of spatial skills, related to damage to the right hemisphere, a correlation first recognized by the British neurologist Hughlings Jackson in 1876).

The asymmetric function of the mammalian brain is a puzzle because no physical difference has been observed between the two hemispheres.

Split-brain research conducted by Roger Sperry and Michael Gazzaniga showed that both hemispheres are capable of performing the same processing.

The US neurologist Lynn Robertson argues that it is a difference in strength rather than in kind.

There is a tiny difference in the way the two hemispheres process early perceptual information, the very first data coming from the senses, and that difference gets amplified as it is processed in the two hemispheres.

Robertson's theory is that we understand the world at different levels.

The world is made of objects that are made of objects. We perceive, recognize and understand objects at all levels of the hierarchy, and we do so in parallel: we do not build a scene from its parts, and we do not split a scene into its parts. Our perception of the world is a cooperative process of both aspects.

We recognize both a scene and its parts, because recognition proceeds simultaneously at different scales.

Perception is performed by "frequency-sensitive detectors" in the brain. The right hemisphere is better at analyzing low-frequency information, such as spatial information (objects), whereas the left hemisphere is better at processing high-frequency information, such as language. This asymmetry generates a distributed representation of the world that may have been advantageous for evolutionary purposes.

This holds for both vision and audition. The right hemisphere is better equipped for high spatial-frequency patterns as well as high sound-frequency patterns. And viceversa.

The reason there are no visible differences between the two hemispheres is that they perform the same processing on the same patterns of information. The difference is subtle: each hemisphere has a bias for a range of frequencies.

Suspended Animation

Besides handedness, there is another feature that sets the human brain apart from other animals.

Human brains are rather unusual in that they require a steady supply of oxygen. A few seconds without oxygen are enough to cause permanent damage to the human brain, if not death itself. On the contrary, most animals can live quite a long time in different kinds of suspended animation. Many mammals can hibernate. Most seeds can survive months or even years before germinating. All these living beings are capable of suspending or slowing down their cellular activity until the environmental conditions become favorable to growth.

Humans are much more vulnerable to changing environmental conditions.

The Life Of Neurons

In 1891 the Spanish anatomist Santiago Ramon y Cajal proves that the nerve cell (the neuron) is the elementary unit of processing in the brain, receiving inputs from other neurons via the dendrites and sending its output to other neurons via the axon. In 1905 Keith Lucas demonstrated that below a certain threshold of stimulation a nerve does not respond to a stimulus and, once the threshold is reached, the nerve continues to respond by the same fixed amount no matter how strong the stimulus is.

A brain is made of neurons (nerve cells) which communicate via junctions called synapses. Neurons are the largest cells in the human body. Neurons are extremely simple units that can be viewed as switches. What creates the complexity of the brain is the synapses that connect the neurons. A human brain has about 100 billion neurons (50% in the cortex), with an average of 10,000 synapses per neuron, which yields about 500 trillion synapses. Neurons die and are born all the time. Synapses are destroyed, created and modified even more rapidly.

The Italian biologist Luigi Galvani originally suggested that nerve cells were conductors of electricity already in 1771, but it wasn't until 1924 that the electrical activity of the brain was recorded (by Hans Berger, the first electroencephalograms). In 1921 Otto Loewi demonstrated chemical transmission of nerve impulses, proving that nerves can excite muscles via chemical reactions (notably acetylcholine) and not just electricity. In the 1940s the Australian neurologist John Eccles, the German neurologist Bernard Katz and others clarified how neurons communicate via chemicals, generally referred to as "neurotransmitters" (originally discovered in 1921 by the Austrian pharmacologist Otto Loewi).

Under appropriate conditions, a neuron emits an "action potential", which a synapse converts into a neurotransmitter and sends to other neurons. More precisely, neurotransmitters are synthesized by the cell's body and stored in the synapses until a nerve pulse is generated. This chemical messenger can either excite (start firing an action potential of its own) or inhibit (stop firing the action potential) each receiving neuron. Neurons are binary machines: either they fire, or they don't. If they fire, they release always the same amount of neurotransmitter. Each neuron can synthesize and therefore release only one kind of neurotransmitter. There are about fifty kinds. Each neurotransmitter has a particular effect on receiving neurons and can therefore yield a different "pathway" within the brain.

Neurotransmitters do matter. Each seems to have a different kind of message. For example, endorphins are related to pain. Each neuron only releases one kind of chemical (which can contain more than one neurotransmitter, but is always the same combination). A neurotransmitter can be received only by appropriate receptors. De facto, each neurotransmitter creates a sub-net of neurons. Each seems to contribute to different aspects of mental life.

A particular class of neurotransmitters, the "neuromodulators" (dopamine, serotonin, acetylcholine, etc), spread through large areas of the brain and are crucial to determine the large-scale dynamics of the cortex.

The "intelligence" of the brain is due to the high number of connections, which cause a simple signal to generate a very complex chain reaction of activation of neurons around the brain.

This intelligence would be pointless if it didn't relate to the rest of the body. The nervous system extends throughout the body via nerve fibers. There are two kinds of nerve fibers and they are both one-way only: "afferent" nerves connect the senses to the brain, and "efferent" nerves connect the brain to the muscles.

The Biorhythms

The nervous system is made of two main subdivisions: the central nervous system (the brain and the spinal cord) and the peripheral nervous system (in particular the autonomic nervous system that controls the heartbeat, breathing and other bodily functions).

This complex apparatus relies on a number of internal clocks: heartbeat (approximately one per second), breathing (approximately once every 4 seconds), REM sleep (4 or 5 times per night, at approximately 90 minute intervals), sleep/wake or "circadian" (every 24 hours), menstruation (every 28 days), hibernation (every 365 days), the thalamus' rhythm (40 times a second), the amygdala's rhythm, etc.

All these "biorhythms" are registered in the brain, although they cannot be consciously perceived. The synchronization of such a complex system of biorhythms is accomplished by the brain.

In 1972 the US neurologists Robert Moore and Irving Zucker discovered that the suprachiasmatic nucleus (at the base of the hypothalamus), a cluster of about 10,000 neurons, keeps the central clock of the brain, the "circadian" clock, that dictates the day-night cycle of activity. The cells of the suprachiasmatic nucleus perform chemical reactions that take about 24 hours to complete. Those cells are connected to other regions of the brain and the products of their chemical reactions directly affect the activity of those regions.

Triggered by the suprachiasmatic nucleus, melatonin secretion starts after sunset, induces sleep and lowers the temperature of the body. Blood pressure starts to rise with sunrise. Then melatonin secretion stops and we wake up. We become more and more alert, as both blood pressure and body temperature increase. At sunset the cycle resumes.

Circadian rhythms are so common among species (even plants) that they may be one of the oldest attributes of life.

The importance of these clocks cannot be overlooked. The behavior of living organisms changes as the day progresses, because their clocks tell them so. It is not consciousness that tells us what to do: it is our inner clock that dictates much of our behavior. The brain is slave to its clock.

A circadian clock is actually present in every cell of the body: an isolated cell in the laboratory still follows a 24-hour cycle. What the suprachiasmatic nucleus does is to synchronize the 24-hour cycles of all the cells in the body.

We are controlled by several different clocks rather than by just one clock, a fact that appears to be a senseless complication. Part of the complexity of the brain may be due precisely to the need to "transduce" each of these rhythms into the other ones, otherwise different organs could not cooperate.

The operations performed by the various organs of the body occur in continuous quantities, not discrete quantities. For example, we can drink any amount of water (not only some set amounts) and exhale any amount of air. However, the "functioning" of those organs is discrete, not continuous: a clock sets their rhythm.

Somehow the body needs to "pace" each of its many internal functions. The clocks may exist precisely because they enable synchronization among wildly different organs that happen to depend on each other.

A number of biological clocks (many more than we have discovered so far) may be presiding over all the vital activities of the body. In other words, "life" might just be a store of biological clocks. Their rhythms "are" our lives". These clocks activate programs that keep repeating the same functions at fixed intervals. As they operate and interact with many other such repetitive programs, we live, we act, we behave. Our behavior may simply be the outcome of those numerous programs repeating their mechanical actions. Besides the obvious ones for breathing, heartbeat, seeing, and so forth, there might be repetitive programs for detecting this or that feature of the world, for scanning memory, for learning new knowledge, and so forth, all of them running periodically in the brain over the available information.

The Endocrine System

The endocrine system includes all of the glands that regulate the functions of organs through the secretion of hormones, thereby maintaining the body's homeostasis. The hypothalamus secretes several hormones. The pineal gland (behind the thalamus) produces the hormone melatonin. Glands throughout the body, from the thyroid gland to the ovaries, produce dozens of hormones. Hormones spread via the bloodstream to the entire body. Basically the body has two control pathways that stretch from the brain to the toes: the nervous system and the bloodstream. The nervous system provides fast and targeted signaling, whereas the endocrine system provides slow and widespread signaling.

The pituitary gland, located in the hypothalamus, connects the nervous system and the endocrine system

Brain Waves

Then there are the "brain waves" measured by electro-encephalograms. The frequency of 13-30 Hz ("beta waves") prevails in mental states of concentration, studying, stress (your brain is producing beta waves as you are reading this). Alpha waves (8-13 Hz are typical of states of relaxation and meditation. Theta waves (4-8 Hz) emerge in deep states of meditation, cases of religious ecstasy, and REM sleep. Young children are in theta most of the time. Delta waves (up to 4 Hz) characterize deep sleep. Babies are in delta most of the time. There are also gamma waves (25 to 100 Hz, but mostly 40 Hz) and mu waves (8-13 Hz, same band as the alpha waves, but the mu waves are specialized for the motor cortex).

A sine wave (or sinusoidal oscillation) is determined by three features: amplitude (the energy of each cycle, the difference between crest and rest), frequency (the number of cycles per second, which, given the velocity of propagation, also determines the wavelength, i.e. the distance from crest to crest) and phase. Signals are synchronous when they have the same phase (each cycle of the wave begins and ends at the same time). Coherence measures the phase consistency between two signals. Coherence is one, for example, when the two signals are perfectly synchronous. A signal is generally a combination of several frequencies, each with its own amplitude and phase. The fast Fourier transform is the mathematical technique that separates the various frequency components of a signal. A signal can be synchronous/coherent with another signal at some frequency but not at others. Traditionally, the electroencephalogram measures large-scale electrical activity in the brain. Depending on the mental and bodily state, the "signal" picked up by an electroencephalogram contains different frequencies. When the local electrical activity is measured, instead, local differences become evident.

The alpha, beta, delta, theta and gamma brain waves are the result of measurements of temporal patterns. Spatial analysis (locating the source or the distribution of a "wave") has traditionally been more difficult.

The idea that the frequency components of the electroencephalogram have a specific function, and the application of mathematical analysis to the electric activity of the cerebral cortex, dates back to the Russian psychologist Mikhail Livanov, who organized a symposium in 1964 on "Mathematical Analysis of the Electrical Activity of the Brain". He speculated that the spatial distribution of these different frequencies through the cerebral cortex played an important role in our cognitive life. While studying rabbits, he observed that the electrical patterns in the motor and visual regions of the brain were synchronous when the rabbit was reacting to a visual stimulus, and concluded that functional correlation between two brain regions manifests itself in the synchrony between the corresponding brain waves.

Perception And Sensation

A sensory input undergoes a long series of transformations before leading to a "sensation". Its features are extracted as it is channeled through the brain. Information about its location is isolated early in this process and kept separate from the features. Different regions of cells in the cortex process different features. As the sensory information travels through these regions, each region extracts some feature.

For example, one of the most complex sensory inputs is the one that comes from the retina. The component about the location is isolated and routed elsewhere, while information about the sensory input (e.g., the texture, size, color, shape of an image) is still traveling together in one chunk (this produces, for example, the sensation and emotion of the whole image). The location component follows, instead, a completely different path. "What" and "where" are processed separately and in parallel.

Concepts And Memories

Last but not least, the brain "categorizes".

The brain needs to categorize environmental stimuli that are valuable for survival. Every second we are bombarded with millions of sensory stimuli and we slowly force an organization on them, discarding some and retaining others. We turn a chaotic deluge of random stimuli into an ordered flow of patterns. We categorize them in "events", "situations" and "things". The newborn is simply powerless in the face of excess information. We learn slowly to process information and reduce its complexity by organizing our brain.

The brain understands what matters by reducing stimuli to concepts, categories. The brain remembers by linking new memories to old memories.

The brain, in the face of huge daily sensory stimulation, has a crucial task:
- understand what matters
- understand what does not matter
- remember what will still matter
- forget what will never matter again

Memories are stored in neural activity patterns which are distributed throughout the brain. This system of storage is more versatile and redundant than a "container". It is more redundant because a portion of the brain may be damaged without seriously affecting the overall pattern representing a memory. It is more versatile because it makes it easier to link many different memories.

Retrieving a specific memory does not entail finding its location, but turning on its pattern of neural activity.

The Dynamic Brain

The British zoologist John Young can be credited with starting (in 1964) "selectionist" thinking about the brain. He understood that learning could be the result of the elimination of neural connections (or, better, weakening of synapses).

In the 1950s Roger Sperry demonstrated that the brain is pre-wired by the genetic program to deal with some categories and to coordinate some movements. Sperry proved that experience is not enough to shape the brain, whereas Young proved that experience shapes the brain in a Darwinian manner.

The essential feature of the brain is that it is a dynamic system, capable of changing very quickly. Even the adult brain "grows".

Only about a quarter of the brain is already grown at birth. It is not only quantity, it is mainly quality that is missing. In fact, quantity is taken care of very rapidly: four weeks after conception, an embryo is creating neurons at the fantastic rate of about 500,000 per minute. Six weeks after conception (three months before being born), a fetus has actually more neurons than it will ever have. "Intelligence", though, comes from the synapses.

At birth, the brain has fewer synapses than an adult brain. While the brain comes with some synapses pre-wired, many are formed in response to the environment. Synapses proliferate rapidly during the first two years of life. Virtually every event of a child's life leads to the creation of synapses. By far, it is the cortex that witnesses the highest frequency of synaptic creation, while the rest of the brain is largely unchanged after birth. At birth a human brain is much less "finished" than the brains of other animals.

Concurrent with the explosion of synapses is a rapid pruning away of those that do not get used. The brain is built through the interplay of genes and experience. The newborn brain comes equipped with a set of genetically based rules that specify how learning takes place. Then the brain is literally shaped by experience (by what is used and what is not used).

The infant's brain organizes itself under the influence of waves of "trophic" factors. Such factors are chemicals that promote the growth and interconnections of nerve cells. They are released in waves so that different regions of the brain become connected sequentially. Again, the process is modulated by experience (by what happens to the infant, i.e. by which stimuli enter the infant's brain).

Besides the creation and deletion of synapses, the brain undergoes another phenomenon that shapes its ability to "think": synapses change, again, in response to the environment. Synapses are not simple links between neurons, they are more or less effective in implementing such a link.

Donald Hebb's hypothesis, formulated in the late 1940s, is that the basis for neural development lies in a selective strengthening or inhibition of synapses. Synapses that get used are reinforced, while synapses that are not used are inhibited. This dual process molds the structure of the brain in a Darwinian fashion: the more "useful" synapses are the ones that survive. These synaptic changes are the basis for all learning and memory.

Hebb had already realized that metabolic change occurs in the brain all the time.

The selective strengthening of the synapses causes the brain to organize itself into "cell assemblies", regions of interconnected self-reinforcing subnets of neurons that form for long periods of time. These acts of "reinforcing" are more than mere "stimulus-response" pairs: they are reverberating processes that occur over a network of cells. Each assembly represents a fragment of a concept. An assembly may overlap others, so that concepts are naturally linked into larger concepts. Each resonating cell assembly behaves like a rule: triggered by an event, it will fire for a while at a higher rate.

Psychological conditioning is ubiquitous in animals because it is a property of the elementary constituents of the brain.

Another of Hebb's great intuitions was the "phase sequence". A cell assembly facilitates the formation of another one, normally in conjunction with an external stimulus. A series of chained cell assemblies constitutes a "phase sequence", in which, basically, one thought leads to another.

The elementary constituent of "thought" is actually a population of interconnected neurons (a cell assembly) rather than the individual neuron.

The brain is an evolutionary system: genes determine only its initial configuration, whereas experience molds the brain according to Darwinian principles of selection.

Selection Processes Of The Brain

Darwinian thinking emphasizes selection over instruction. Some variations are "selected" by the environment over others. Adaptation to the environment is a process of selection. Selection processes are ubiquitous in nature.

In 1955 the Danish immunologist Niels Jerne ("The natural selection theory of antibody formation") discovered that a selection process also presides over the immune system. The traditional view of the immune system was that it is capable of manufacturing protein molecules, or "antibodies", in order to neutralize foreign antigens (viruses, bacteria, etc). Jerne and Gerald Edelman discovered that, on the contrary, the immune system routinely manufactures all the antibodies it knows how to make (the Austrian physician Karl Landsteiner had already demonstrated this). Whenever the body is attacked by foreign antigens, some antibodies (the ones that best "bind" with the invader) are selected and start multiplying rapidly to cope with the invasion.

Antibodies are created by the thousands even "before" the body is attacked by anything. An invasion results in a rapid increase in the rate of production of the one antibody that matches the intruder. In a sense, it is the intruder (not the immune system) that decides which antibodies need to multiply.

In 1968 Jerne wondered whether a selection process could also account for the mind: do we learn new concepts or are useful concepts chosen by the environment among a pre-existing array of concepts? Do we create a plan of action or is an action selected by the environment from a pre-existing set of actions? Do we think or is a thought selected by the circumstances from a vast pool of possible thoughts?

Do we design our mental life, or is our mental life a continuous process of environmental selection of concepts in our brain?

Does our mind manufacture ordered thought or does it manufacture chaotic mental events that the environment orders into thought?

Socrates believed that all learning consists in being reminded of what we already know. Jerne, basically, updated Socrates' idea to Darwinian thinking: every being is equipped with a library of all possible behavior and cognitive life simply consists in finding (within that library) the behavior that best copes with the environmental conditions.

The genes encode that "library". They encode information accrued over millions of years of evolution.

The mind already knows the solution to all the problems that can occur in the environment in which it evolved over millions of years. Given a problem, it is only a matter of retrieving the appropriate solution. Indirectly, it is the environment that selects what the mind does. And it is the genes that have restricted what the possibilities are for the mind. And the genes too were selected by the environment.

Further similarities between the immune system and the neural system were discovered in the following decades. For examples, in 1995 the US pediatrician Abraham Kupfer showed that immune cells communicate information via synapses in a manner similar to how neural cells communicate information. The information they exchange is presumably about the present dangers within the body.

Neural Darwinism

Drawing from Jerne's ideas, in the 1970s the US biologist Gerald Edelman applied the "selectional" theory of the immune system to the brain. His "Neural Darwinism" is a selectional theory of brain development.

Edelman was after a rational explanation for two apparently bizarre facts. First, there is no way that the human genome can specify the whole complex structure of the brain. Second, individual brains are wildly diverse. One would instinctively expect the opposite: all information about the brain should be encoded in DNA and every individual should get pretty much the same brain.

Edelman was aware that, before birth, the genetic instructions in each organism provide general constraints for neural development, but they cannot specify the exact location and configuration of each cell. After birth, innate values, i.e. "adaptive cues" (such as "looking for food"), generate behavior and therefore feedback from the environment, which in turns helps "select" the neural configurations that are more suitable for survival. During this on-going process of "learning", the brain develops categories by selectively strengthening or weakening connections between "neural groups". Individual experience "selects" one configuration of neural groups out of all the configurations that are possible. Note that the unit that gets selected is not the individual neuron but a neuronal group.

Edelman's neural groups are a variation on the "cortical columns" of the cortex analyzed by the US neurologist Vernon Mountcastle in 1957 ("Modality And Topographic Properties Of Single Neurons Of Cat's Somatic Sensory Cortex"). Mountcastle proved that neurons (that perform like functions) are not only organized in horizontal layers, but also in vertical columns. He revealed the modular organization of the brain.

Edelman views the functioning of the brain as resulting from a morphological selection of neural groups. Neural groups "compete" to respond to environmental stimuli. That is why each brain is different: its ultimate configuration depends on the stimuli that it encounters during its development.

"Adhesion" molecules determine the initial structure of neural groups, the "primary repertory". Experience determines the secondary repertory. Repertories are organized in "maps", each map having a specific neural function. A map is a set of neurons in the brain that has a number of links to a set of receptor cells or to other maps.

Maps communicate through parallel bidirectional channels, i.e. through "reentrant" signaling. Reentry is not just feedback because there can be many parallel pathways operating simultaneously. The process of reentrant signaling allows a perceptual categorization of the world, i.e. to relate independent stimuli. This feature enables higher level functions such as memory.

Categorization is a process of establishing a relationship between neural maps (through that reentry mechanism). Categories (perceptual categories, such as "red" or "tall") do not exist physically. They are not located anywhere in the brain. Categories are that (on-going) process.

Basically, Edelman believes that neural groups are bound to compete and evolve in a Darwinian way, and eventually self-organize as neural maps (purposeful assemblies of neural groups).

A further level of organization leads to (pre-linguistic) conceptualization. Conceptualization consists in constructing maps of the brain's own activity, or maps of maps. This process of "global mapping" indirectly retains knowledge of past activity. A concept is not a thing. It is a process. The meaning of something is an on-going, ever-changing process.

According to Edelman's view, brain processes are dynamic and stochastic, whereas the traditional view held the brain to be static and deterministic. Furthermore, the brain is not an "instructional" system but a "selectional" system. It evolves not by changes in a constant set of neurons but by selection of the most valuable neural groups among those that were created at birth. And the elementary unit of this process is not the single neuron, but the neural group.

This Darwinian model of the brain explains the non-linearity between the complexity of the genome and that of the brain. The brain is not a direct product of the information contained in the genome. It uses much more information than is available in the genome, i.e. information derived from experience, i.e. indirectly received from the environment.

A Brain in Transition

There are slight variations on the idea that the brain is a dynamic system.

The French neurobiologist Jean-Pierre Changeux introduced the paradigm of "epigenesis by selective stabilization of synapses". In his model too the nervous system makes very large numbers of random multiple connections while, at the same time, external stimuli cause differential elimination of some connections (useful ones are retained, useless ones are eliminated). Phenotypic variability (differences among individual brains) is the result of experience. As he put it, "the Darwinism of synapses replaces the Darwinism of genes".

Interestingly, he noticed that phenotypic variability increases with the increase in brain complexity (the simpler the brain of a species the more similar the brains of individual members of that species). The evolutionary advantage of the human species stems from the individual, epigenetic variability in the organization of neurons, which resulted in greater plasticity in adapting to the environment.

The US neurologist Dale Purves, for example, has shown how brain cells are in a perennial state of flux, creating and destroying synapses all the time. Neural activity caused by external stimuli is responsible for the continual growth of the brain, and for sculpting a unique brain anatomy in every individual based on the individual's experience.

The British neurologist Semir Zeki argues that perception and comprehension of the world occur simultaneously thanks to reentrant (reciprocal) connections between all the specialized areas of the cerebral cortex. The function of the sensory parts of the cortex is to categorize environmental stimuli. The brain copes with a continually changing environment by focusing on a few unchanging characteristics of objects out of the countless ever-changing bits of information that it receives from those objects. The brain cannot simply absorb information from the environment. It must process it to extract those constant features that represent the physical essence of objects. The brain is basically programmed to make itself as independent as possible from world changes.

Patterns And Brains
The US mathematician Ben Goertzel believes that thinking, like life, is a process of evolution by natural selection. In general, Goertzel believes that Darwinism must be supplemented with a theory based on self-organization of complex systems. An organism that, coupled with the other organisms in its environment, generates a large amount of emergent pattern is more likely to survive. Consequently, his model replaces Edelman's neural maps with hierarchical structures that generate emergent pattern. Then neural maps can be viewed as populations that are reproducing sexually and evolving by natural selection. Basically, brain regions are equivalent to ecosystems. And Stephen Jay Gould's punctuated equilibrium applies as well to the cognitive development of an individual.

Goertzel views minds as sets of patterns interested in recognizing, creating and executing patterns. A mind recognizes patterns in the world, matches them to patterns that are contained within itself, and then creates new patterns both in the world and within itself.

The US physicist Eric Baum argues that mind originates from an "Occam program", a program that stores only the information that is truly needed and in a minimal form. Baum argues that the brain is an unlikely candidate for such a program. The genome, on the other hand, is just that: a program. Baum thus views the genome as the software (or, better, the source code) and the brain as the hardware (or, better, the executable code) that implements his Occam program to deal with environmental patterns, translate them into minimal mind patterns and then enact them as efficient behavioral patterns. And evolution is the software engineer that wrote the program.

The US neurobiologist Walter Freeman pioneered neurodynamics when, aiming to explain the meaning of an electroencephalogram, he introduced the concept of "mass action", the "force" that large populations of neurons in the cortex generate by synchronizing their firing of action potentials. This "force" is responsible for bursts of cortical activity that resembles the vortices of tornadoes and hurricanes. Freeman suggested that these "bursts" may correspond with the formation of percepts. Freeman viewed these bursts of neural activity as the moments in time when the brain binds sensory inputs with memories. Cortical neurons belong to "sets" whose internal behavior can be modeled as made of three components. Linear dynamic equations can express two of them: the oscillation in time and the oscillation in space. Alas, when one adds the third component (the massive interconnections and feedback mechanisms of the neurons), the result is a system of nonlinear partial differential equations in time and space (Freeman thought that the "chaos" generated by these equations is precisely what makes consciousness possible).

These scientists place different emphasis on "how" the mind decides to build patterns. Does it use a goal-oriented approach (makes predictions that are useful for its goals), does it use a genetic-oriented approach (makes predictions that match its genetic repertory), or does it use a computational approach (makes the predictions that reduce the complexity of the world)?

The implementation in the brain of this prediction machine is the link between neural processes and symbolic processing: neural processes ultimately constitute the vehicle to create and manipulate symbols.

Presumably, this function involves a massive use of some form of Hebbian learning, leading from disjointed instances to more and more organic and abstract representations.

Thus generalization and metaphorical thinking are the fundamental basis of cognition.

The Selectional Mind

The US neurophysiologist Michael Gazzaniga extended Jerne's ideas to prove that a selection process also governs higher mental functions such as language and reasoning.

Gazzaniga agrees with Edelman that, during growth, selection processes determine how a brain is wired for adult functioning. Brains are born with a vast number of pre-wired circuits, which nonetheless offer many alternative options for development; and experience determines which of these pre-existing brain circuits are used. Many possible connections can be made, but only some are selected by experience.

The mind is shaped by the environment; but the environment can only shape it as far as genetically-fixed parameters allow. It is more appropriate to say that the environment "selects" from the possible outcomes.

Neurons exist because it was written in the genetic code. They perform their function no matter what. It is the interaction with the environment that will prefer some neurons over others. But, ultimately, the neurons were already there.

Gazzaniga differs from Edelman in emphasizing the importance of innate structures. Learning consists in discovering already built-in capabilities. The phenomenal rate of learning in children can be explained by admitting that children already "know". What they are learning is what is selected through interaction with the environment. Noam Chomsky's universal grammar is an example.

Children quickly learn a language because linguistic knowledge is present in their brain at birth and all their brains have to do is pick what is consistent with the specific language spoken around them.

All humans are equipped from birth with some general features that allow for intelligent behavior in our world. Experience (i.e., interaction with the environment) will decide how that behavior will materialize.

Convergence Zones

A new paradigm was introduced in the 1980s by the Portuguese biologist Antonio Damasio.

When an image enters the brain via the visual cortex, it is channeled through "convergence zones" in the brain until it is identified. Each convergence zone handles a category of objects (faces, animals, trees, etc.) A convergence zone does not store permanent memories of words and concepts but helps reconstructing them. A convergence zone is not a "store" of information, but an "agent" capable of decoding a signal (of reconstructing information). In this function, they resemble an "index" that can be used to organize a perception.

There is no specialized region of the brain that encodes an event (a memory). The various features of a perception are held in the places where they were analyzed (somewhere in the cortex). The convergence zones are different regions of brain that manage the task of connecting those fragments of perceptions and of connecting them to previous "memories". Convergent zones also produce output. If convergence zones reactivate simultaneously fragments that used to be connected when they were first "memorized", then we "remember" the event represented by the set of those fragments.

Once an image has been identified, an acoustic pattern corresponding to the image is constructed by another area of the brain. Finally an "articulatory" pattern is constructed so that the word that the image represents can be spoken. There are about twenty known categories that the brain uses to organize knowledge: fruits/vegetables, plants, animals, body parts, colors, numbers, letters, nouns, verbs, proper names, faces, facial expressions, emotions, sounds.

Convergence zones exist at several levels. A convergence zone may be responsible for linking the attributes of a face, while another may be responsible for linking the face to other concepts or faces.

Convergence zones form a hierarchy of specialized agents (although they are connected in a network-like fashion). Each convergence zone is the focal point for the integration of disparate features. Convergence zones "bind" together objects, concepts and events at different levels of cognition.

Convergence zones behave like indexes that draw information from other areas of the brain. The memory of something is stored in bits at the back of the brain (near the gateways of the senses): features are recognized and combined and an index of these features is formed and stored. When the brain needs to bring back the memory of something, it will follow the instructions in that index, recover all the features and link them to other associated categories. As information is processed, moving from station to station through the brain, each station creates new connections reaching back to the earlier levels of processing. Convergence zones enable the brain to work in reverse.

Phantoms in the Brain

By analyzing different kinds of brain damage, and the feelings associated with phantom limbs (people with missing limbs can still feel pain in those non-existent limbs), the Indian neurologist Vilayanur Ramachandran concluded that the brain constructs cognitive maps that are, basically, plausible interpretations of the world. It is those maps that cause all mental life, starting from perception itself. For example, the limb is no longer there, but its representation in the brain is still there, and thus the person feels it as if it were still there. Whether it is truly there or not is negligible compared with the fact that it is represented in the brain. By generalization all mental life could be "phantom", because that is a general behavior of the brain.

All sensory experience is an illusion. All feelings are illusions. Even the self consists of an illusion, largely constructed out of interactions with others. The brain creates these representations of different kinds (from representations of limbs to representations of the i) and then believes that they truly exist and they get associated with feelings. (Thus the solution to the pain caused by a phantom limb would be to induce the brain to believe that the phantom limb does not exist anymore, i.e. to remove the representation of that limb in the brain).

In a sense, the entire body is a "phantom limb": the brain constructs its existence and then "feels" it.

Mirror Neurons

The Italian neurologist Giacomo Rizzolatti discovered ("Action Recognition in the Premotor Cortex", 1996) that the brain of primates uses "mirror" neurons to represent what others are doing. De facto, my brain contains a representation of what someone else is doing, and that representation helps me "understand" what the other person is doing, for example her intention and her emotions. We effortlessly understand the intention and emotion of others not because we carry out complex reasoning procedures about their actions but because their intentions and emotions are physically reproduced inside our own brain. In fact, a brain only needs to see the beginning of an action by another person in order to guess the intention of it: based on the context, the mirror neurons instantly reproduce the brain state of the other person and therefore help to understand what the other person is trying to do and what will happen next. These mirror neurons are widespread in the cortex of primates (not only of humans). These mirror neurons fire both when the action is performed and when the action is observed in other individuals. Ramachandran ("Mirror Neurons and Imitation Learning", 2000) subsequently speculated that mirror neurons may be crucial in learning and understanding language.

Because mirror neurons fire not only when an action if performed but also when observing someone else performing that action, they allow us to understand the intentions of someone else's action and to empathize. They de facto simulate what others are doing. Ramachandran credits mirror neurons with shifting the main driver of human evolution from the genome to culture. As culture became more and more important, evolution started selecting the brains that had the best mirror neurons.

Similarly, the human brain has "canonical" neurons that fire both when we perform an action on an object and when we simply see that object. For example, some canonical neurons fire both when we grasp and object and when we see that object (in other words, they seem to fire in response to the property of "graspability", regardless of whether we actually grasp the object or not).

The brain is prewired for understanding motives and for deceit. Babies understand (or at least try to guess) the motives of the people around them way before they can understand the language. When someone asks "do you know the time", we expect the answer to be either "no" or the current time. The answer "yes" is a joke. If you ask for the directions to the library at a time when the library is closed, you are likely to be told "It is closed" and not how to get there. If you pull out your camera in a museum where photography is not allowed, a security guard is likely to tell you "No photos". A lot of what we do in society is driven by our understanding of other people's motives. It is a fundamental feature of the human brain that we continuously build theories of other minds.

Synesthesia

The US neurologist Richard Cytowic speculated that synesthesia was the normal state in the primitive human brain, when the senses had not fully separated.

Ramachandran, instead, advanced the hypothesis that the cause is interference between regions of the brain specialized in different tasks; and argued that cross-activation between specialized modules is pervasive and, ultimately, accounts for the kind of inference and abstraction at which human brains are best, and even for human creativity.

Neural Communication

There are indications that communication within the brain may involve a different kind of paradigm from the one employed by human-built machines. We tend to think of communication between parts as a signal that travels from a point in the system to another point in the system. In the brain, however, messages do not seem to carry meaning (per se). The Brazilian neurologist Sidarta Ribeiro ("Global Forebrain Dynamics Predict Rat Behavioral States And Their Transitions", 2004) argued that communication is carried out via patterns of activity that involve entire regions. It is the pattern of activity that carries and delivers meaning.

By the same token, the Indian neurologist Mayank Mehta ("Role of Rhythms in Facilitating Short-term Memory", 2005) thought that the neocortex communicates with the hippocampus for the transfer of long-term memories via synchronized patterns of activity (a rhythmic pattern of the excitatory cells in the neocortex corresponds to a rhythmic pattern of the inhibitory neurons in the hippocampus). Different regions of the brain communicate by synchronized firing of neurons (as opposed to a linear transmission of signals from neuron to neuron to neuron).

Reading Minds

The British neurologist John-Dylan Haynes speculated ("Reading Hidden Intentions in the Human Brain", 2007) that one can "read" the thoughts of a brain by analyzing the pattern of neural activity. Each thought is uniquely associated to a pattern, so in a sense that pattern represents the "fingerprint" of that thought. (visual imagery, emotions and plans, including concealed intentions). Since there is a direct correlation between a "train of thoughts" and a decision, once the pattern has been identified one can even predict which decision a person will make.

The Free Will of the Brain

The traditional view of the brain is that, fundamentally, it serves the purpose of "reacting" to what happens in the environment. As the body encounters new situations, the brain decides what the body must do to cope with them. The traditional view is that the brain is activated by the sensorial data and in turns activates the external organs of the body to generate movement.

This view was challenged by the Colombian neurologist Rodolfo Llinas.

First of all, Llinas does not believe that the neuron is simply a switch: Llinas ascribes a personality to each single neuron. They don't simply react to stimuli, they are active all the time, generating patterns of behavior all the time.

Second, Llinas considers the brain a "prediction machine". Organisms that need to move also need to represent the world and make predictions on what is going to happen. Therefore they need a brain. However, his opinion is that, once endowed with a brain, an organism has only limited control of it.

Neurons are always active, even when there are no inputs from the external world. Neurons operate at their own pace, regardless of the pace of information coming in from the outside. A sort of rhythmic system controls their work. They produce a repertory of possible actions. The circumstances "select" which specific action is enacted. For example, the motion of cerebellum neurons results in body movements if the conditions are appropriate (the cerebellum is the part of the brain that controls movement). But, in a sense, the neurons are telling the body to move even when the body is not moving and before the body started moving. Movement is not reactive: it is active and automatic. The environment, in a sense, selects which movement the body will actually perform, but at that point in time the brain may have been ready to perform many other movements.

"I" am a consequence of my brain thinking. My brain is thinking and the environment is deciding what it is thinking, and "I" only exist after the fact (my mental life only exists after the fact, my conscious I, my illusion of being a being, only exists after the fact). I don't think, I simply have the illusion of thinking. In reality the brain is thinking independently from my will, its thoughts shaped by experience, and then, after they have been thought and selected by the environment, I can experience them. Ultimately, the environment is thinking my thoughts!

The Advent of the Brain: Encephalization

The history of the brain is the history of the nervous system. Multi-cellular organisms eventually developed the ability to control their cells. Each cell had its own internal mechanism of control, and somehow was capable of mediating with the other cells. The nervous system is made of cells that mediate the need of the cells of the body.

This function was evolutionarily useful and therefore persisted and evolved. It presumably evolved both in quantity and in quality: more and more nervous tissues would coordinate the movements of the organism, and more and more processing would be performed based on performance.

Eventually the nervous system started building abstractions of controls, the equivalent of "representing" the body and its interaction with the environment. At this point it made sense that the nervous system became "headquartered" in one specific place, rather than being spread throughout the body.

"Encephalization" is the name given by the British neurologist John Hughlings Jackson to the process whereby the nervous system of living organisms grew in size and importance especially in the head. What used to be a distributed system of control then became a centralized system of control. In mammalians, more and more centralized tasks were created via the newly born cerebral cortex.

The modern brain was born.

Multiple Brains: the Advent of Cognition

Because higher functions of the brain tend to be generated by the regions (such as the cortex) that appeared in more recent species (such as us), it is likely that the human brain has accumulated functions and structures over the ages. Hughlings Jackson also noticed that a loss of brain function is often compensated by a gain in another brain function, and concluded that an evolutionarily "older" brain takes over whenever the "newer" brain is disabled: the older brain is still there, even though a newer brain grew on top of it.

Today's brain basically "summarizes" its evolutionary history: its structure and functioning contain its predecessors.

Older creatures tend to have no central nervous system, but rather a loose affiliation of nerve fibers. As we move down the genealogical tree, that chaotic form of communication among cells gets disciplined through a more and more centralized system that performs more and more sophisticated processing of the signals. Hot-blooded animals also need to control temperature and require a more complex control mechanism. Earlier mammals exhibit a forebrain and later mammals developed the cerebral hemispheres. Throughout this evolution of more and more refined nervous systems, the earlier ones remained around. The primitive forebrain is still part of the human brain (and it accounts for a lot of our emotional life). Loose networks of nerve fibers still control organs around the body, and often the brain cannot override them. And so forth. One can recognize within the human brain the facsimile brains of amoebas, insects, worms, etc.

The US biologist Philip Lieberman proposed that the brain consists of a set of specialized circuits that evolved independently at different times. Many specialized units work together in different circuits (the same unit can work in many circuits). The overall circuitry reflects the evolutionary history of the brain, with units that adapted to serve a different purpose from their original one. For example, rapid vocal communication (as in "speaking) is actually responsible for the evolution of the human brain, and not the other way around.

Lieberman's "circuit model" was derived from the model of the brain worked out by the US physician Norman Geschwind ("Disconnexion Syndromes In Animals And Man", 1965, the manifesto of behavioral neurology) in order to explain aphasia, a model that reconciled localization and connectionism.

Besides language, another unique trait of the human race (and therefore of the human brain) is the moral code, in particular altruism. This would also be a relatively recent development, and presupposes circuitry for language and cognition.

Microgenesis
"Microgenesis" is an extreme version of this view.

The idea, originally advanced by the US psychologist Jason Brown, is that mental process recapitulates evolutionary process.

Microgenesis assumes that the structure of perceptions, concepts and actions (and mental states in general) is not based on representations but on processing stages that last over a micro-time, propagate "bottom-up", and are not conscious. A representation is but a section of a processing continuum. Mind is not the final representation, it is the very series of processing stages. Earlier processing stages remain part of the final stage just like a child's early stages of development persist as subconscious themes in the adult's cognitive life.

Microgenesis means that at every point in time the brain revisits the very steps that made it evolve from a simple stimulus-response mechanism to a complex control system. Every single line of reasoning goes through all the layers, starting with the primitive emotional reactions that are common to many animals and ending with the sophisticated logical processing that is unique to humans.

Microgenesis is the equivalent for micro-times of ontogenesis (growth of the individual) and phylogenesis (evolution of species). They are the expression of the same general process over different time scales. Microgenesis is sort of instantaneous evolution.

The theory implies that symptoms of brain damage represent normal stages in the cognitive life at microscopic level. Therefore they can be used to reconstruct cognitive life. For example, Brown used this technique to reconstruct the way language is produced and understood.

The Triune Brain

The US neurologist Paul MacLean ("The triune brain, emotion, and scientific bias", 1970) popularized the notion that the human head contains not one but three brains: a "triune" brain.

Like the layers of an archeological site, each brain corresponds to a different stage of evolution. Each brain is connected to the other two, but each operates individually with a distinct "personality". The neocortex does not control the rest of the brain: all three parts interact, although it is true that the neocortex interacts in a more "cognitive" manner. But the "brain" that interacts in a more "instinctive" manner can be as dominant and even more. And ditto for the "emotional" one.

The oldest of the three brains, the "reptilian" brain, is a system that has changed little from reptiles to mammals and to humans. This "brain" comprises the brain stem and the cerebellum. It is responsible for species-specific behavior: instinctive behavior such as self-preservation and aggression. The cerebellum and the brainstem constitute virtually the entire brain of reptiles. The most basic life-sustaining processes of the body, such as respiration, heartbeat and sleep, are controlled by the brainstem. More precisely, the brainstem is the brain's connection with the autonomic nervous system, the part of the nervous system that regulates functions such as heartbeat, breathing, etc. that do not require conscious control. It is always active, even when we sleep. It endlessly repeats the same patterns over and over again, mechanically. It does not change, it does not learn. In ancient species this system was basically most of the brain, and limbs and organs were controlled locally.

Most mammals share with us the limbic system, which MacLean believes was born after the reptilian system and was simply added to it. The earliest mammals had a brain that was basically the reptilian brain plus the limbic system. MacLean therefore believes this to be the old mammalian (or "paleo-mammalian") brain. The limbic system contains the hippocampus, the thalamus and the amygdala, which are considered responsible for emotions and emotional instincts (behaviors related to food, sex and competition). These emotions are functional to the survival of the individual and of the species. This system is capable of learning, because it contains "affective" memories, i.e. emotion-laden memories. Ultimately, the limbic system is about "pain" and "pleasure": avoiding pain and repeating pleasure.

The neo-cortex is the main brain of the primates, which are among the latest mammals to appear. All animals have a neo-cortex but only in primates it is so relevant: most animals without a neo-cortex would behave normally. This "neo-mammalian" brain is responsible for higher cognitive functions such as language and reasoning.

The oldest of the three brains is located at the bottom and to the back. The newest sits on top and to the front.

They all complement each other to produce what we consider human behavior. Each is an autonomous unit that could exist without the others.

The elegance of MacLean's model is that it neatly separates mechanical behavior, emotional behavior and rational behavior. It shows how they arose chronologically and, indirectly, for what purpose. And it shows how they coexist and complement each other. They constitute three steps towards human "intelligence".

The US psychologist Anthony Stevens related these three brains to Jung's division of the mind into a conscious, un unconscious and a collective unconscious, the collective unconscious being the oldest one and therefore assigned to the reptilian brain.

An Olfactory Brain

The US neuroscientist Rhawn Joseph argued that the human brain still contains parts that were used by animals that lived hundreds of million of years ago. In other words, we share parts of the brain of many other animals, and, ultimately, one can say that all animals are "linked" by the "collectively shared unconscious" Joseph calls the human body a "living museum" because it contains so many remnants of ancient organs. This is also visible in language: while we have developed sophisticated spoken languages, we still use gestures, that are presumably an archaic form of communication. Old and new languages coexist. We often communicate unconsciously to other beings precisely because we still have, like it or not, the old languages. For example, a facial expression is enough to communicate our state of mind.

Neurons (nerve cells) first appeared 700 million years ago. When neurons got connected, the first brain was born. Joseph believes that the first major grouping of neurons occurred among olfactory cells, that originally may have been external cells. Eventually they migrated inside the body and created an olfactory lobe. Later, a similar fate turned visual cells into the visual lobe. The growth of these two lobes over evolutionary time eventually yielded the brain as we know it (the two hemispheres).

The olfactory lobe also evolved into the limbic lobe, that still controls many of the "instinctive" activities (in both humans and other animals). The cells of the limbic lobe created more and more layers, and eventually created the cortex. Thus the fundamental structure of the modern human brain evolved from the olfactory lobe.

Among the various forms of communication that are crucial to our understanding of the world, Joseph believes that odors play an important role. The nose contains the most exposed (unprotected) neurons of the human body. The mucosa of the nose is directly connected to the hippocampus and the amygdala, which are instrumental in creating memories. It is likely that living beings developed the ability to analyze chemicals (odors) in order to understand changes in the environment and to sense other beings (in fact, our bodies still excrete odor-generating chemicals from the skin). Odors, after all, control sex and aggression and many other basic activities of most species.

The Olfactory Brain

The US psychologist Lucia Jacobs ("From chemotaxis to the cognitive map", 2012) points out that today olfaction for cognition is used by all animals, from nematodes to mammals, and it is, in fact, the only universal sense. It also happens to be the most efficient, even today surpassing vision in accuracy and capacity. Our brain is capable of coding a trillion olfactory stimuli. We can detect a few molecules from the source in an immediate manner that is not matched by vision. Furthermore, for all animals the largest gene family is devoted to olfaction.

She argues that all brains existing today are inherited from a common ancestor. The first brain evolved underwater, in a world defined by chemicals. Single-celled organisms moved in response to chemicals. The first sensory system dealt with a rich chemical world, the marine world. Olfaction was the original sense for remote sensing. More importantly, olfaction helps to map where an animal is in space. Early brains learned to create maps based on turbulent gradients. The olfactory sense needs to be viewed as a mapping system.

Once the brain had acquired that technology, vision was relatively easy to evolve.

Vision did evolve late in evolutionary history. There are no fossil eyes before the Cambrian explosion. Jacobs speculates that the Cambrian explosion was due to the evolution of the ability to map and navigate. That ability created better predators, and it is not a coincidence that, by the end of the Cambrian, the planet is crowded with bloodthirsty armored animals. Vision is now widespread, as are organs for mapping such as antennas. More efficient predators also means a more efficient diet, and that led to larger brains.

Today the olfactory system and the hippocampus are shared by the brains of all vertebrates, and they are the only regions in which new neurons are continuously created. Jacobs argues that the hippocampus is responsible for mapping the territory and it interacts closely with the olfactory system, and the creation of neurons correspond with exploring new spaces. Together these two regions of the brain constitute an integrated olfactory-navigational system.

The Purpose Of A Brain

The British neurologist John Young once argued that "the most important thing about living beings is that they remain alive". A stable state (homeostasis) is what they aim for. Young claims that homeostasis is precisely the job of the brain, the most delicate job of all. Homeostasis requires appropriate responses to the environment, responses whose goal is to keep the state stable. The brain's job is to maintain homeostasis through the selection of appropriate responses. The brain is the computer of a "homeostat". This computer uses a memory, or, better, two memories. One memory is the one created by evolution and inherited at birth, that contains knowledge that can be used in many commonly-occurring situations, whereas the other memory (the one we call "memory") is the one that contains up-to-date knowledge about the outcome of actions during our lifetime. What these two memories do is to collect information (over generations or over a lifetime) that helps the brain define the desired state and maintain that stable state.

The Heat Engine Of The Body

The Spanish neurologist Francisco Mora introduced a simple vision for what brains do for us: they regulate our temperature.

From the beginning life was capable of reacting to temperature: even the earliest unicellular organisms must have been capable of sensing heat and cold.

Heat, after all, was the primeval source of energy for living organisms. These organisms required heat to survive and the main source of heat came from the environment. The progenitors of the living cell, the "protocells", were probably units of energy conversion, converting heat into motion, just like a heat engine. In 1995 the Dutch chemist Anthonie Mueller, the proponent of "thermosynthesis", showed that such systems could form spontaneously in the primordial conditions of the Earth.

If they survived, these organisms must have developed a way to react to positive and negative stimuli, such as correct or excessive amount of heat. The early nervous systems were assemblies made of cells already capable of sensing and reacting to temperature. Proof is that all known organisms, including unicellular ones, are capable of avoiding adverse environmental temperatures. In other words, throughout evolution all organisms were capable of sensing external temperature.

Mora speculates that during the transition from water to land (from stable temperature to wildly variable temperature) the nervous system must have learned to control body temperature. Nocturnal animals must have developed also a means to overcome the loss of environmental heat and produce heat internally. And the autonomic control of temperature was born.

This feature allowed the evolution from cold-blooded animals (animals whose temperature fluctuates with the temperature of the environment, whose only sources of heat are external sources) to warm-blooded animals (animals that maintain constant body temperature by producing heat internally). Cold-blooded animals are dependent on environmental heat: when ambient temperature rises, they are active and seek food; when ambient temperature decreases, their motor activity slows down. Warm-blooded animals overcame this limitation thanks to that self-regulating feature, thanks to the ability of producing heat internally when heat from outside is not enough. And they freed themselves from their habitat: they were capable of changing habitat because they were capable of maintaining their body temperature regardless of changes in the external supply of heat.

A very efficient self-regulating heat engine that maintains constant temperature opens up new opportunities for evolution: one organ that benefited was the brain, that could grow to its actual size and complexity. If it didn't have an adequate supply of energy, the brain would not be capable of performing the tasks it performs. A hot organ is required for thinking.

Modern mammals, who have the highest demand for internally produced heat, regulate temperature through a whole system of thermostats, not just one. Experiments have proved that mammals have not one but many centers of control of body temperature: in the spinal cord, in the brainstem, in the limbic system and mainly in the hypothalamus. Rather than one point of control, this is more like a complex system, that peaks in the hypothalamus.

It is a very accurate system: humans can survive only in a narrow temperature range (a few degrees below or over 37 degrees Celsius a human body becomes a dead body). Why regulate at 37 degrees instead of, say, 20? It turns out that 37 degrees is the ideal temperature for balancing heat production and heat loss.

Ultimately, the brain is responsible for maintaining a constant temperature, the very constant temperature that allows the brain to function.

At the same time, the brain was made possible by a cooling mechanism. Humans are among the few mammals (and the only primates) with a naked skin. Mammals that have a naked skin are usually big: having naked skin helps "cool down" the body and therefore avoid the risk of overheating (overheating would, in particular, damage the brain). Humans, however, cool down by sweating. The US anthropologist Nina Jablonski argued that cooling via sweating enabled the large (heat-producing) brains of humans.

Brain-machine Interfaces

Once understood how parts of the brain work and send commands to the body, it was not difficult to connect brains to machines so that the brain's activity directed the machine's movement.

The first electrical implant in an ear was the work of French surgeons André Djourno and Charles Eyriès in 1957. Building upon their work, in 1961 William House invented the "cochlear implant", an electronic implant that sends signals from the ear directly to the auditory nerve (as opposed to hearing aids that simply amplify the sound in the ear).

Spanish-born neuroscientist José Delgado published the first paper on implanting electrodes into human brains: "Permanent Implantation of Multi-lead Electrodes in the Brain" (1952). In 1965 he famously managed to control a bull via a remote device, injecting fear at will into the beast's brain. He then published his dystopian vision in the book "Physical Control of the Mind - Toward a Psychocivilized Society" (1969). In 1969 he implanted devices in the brain of a monkey and then sent signals in response to the brain's activity, thus creating the first bidirectional brain-machine-brain interface.

In 2000 William Dobelle developed in Portugal an implanted vision system that allows blind people to see outlines of the scene. His patients Jens Naumann and Cheri Robertson became "bionic" celebrities as Dobelle continued to refine his artificial vision system.

In 2002 John Chapin debuted his "roborats", rats whose brains were fed electrical signals via a remote computer to guide their movements.

As for getting data out of the brain into a machine (output neuroprosthetics) in 1998 Philip Kennedy developed a brain implant that could capture the "will" of a paralyzed man (Johnny Ray) to move an arm. In 2005 Cathy Hutchinson, a paralyzed woman, received a brain implant from John Donoghue's team that allowed her to operate a robotic arm.

In 2004 Theodore Berger demonstrated a hippocampal prosthesis that can provide the long-term-memory function lost by a damaged hippocampus. And in 2012 Sam Deadwyler's brain implant managed to even improve the long-term memory of monkeys.

Two-way transmission was just a matter of combining the two technologies.

In 2013 the Brazilian-born neurophysiologist Miguel Nicolelis made two rats communicate (and they were located in two different countries) by capturing the "thoughts" of one rat's brain and sending them to the other rat's brain over the Internet and an electrode.

In 2013 Rajesh Rao and Andrea Stocco devised a way to send a brain signal from Rao's brain to Stocco's hand over the Internet, i.e. Rao made Stocco's hand move, probably the first time that a human was capable of controlling the body part of another human.

In 2014, thanks to Silvestro Micera's team, an amputee, Dennis Aabo, received an artificial hand capable of sending electrical signals to the nervous system so as to create the touch sensation.

When in 2002 the Brazilian-born neurophysiologist Miguel Nicolelis made a monkey's brain control a robot's arm, the monkey's brain was extended in three directions: it expanded the distance at which that brain could operate (because the robot was physically outside the monkey's body), it increased the strength of the force it could generate (because the robot was stronger than the monkey's body), and it reduced the time for a thought to translate into a movement (because the robot reacted faster than the body itself). In 2008 the team made the monkey control a remote robot (in fact, located in another continent) and control it to cause bipedal locomotion. This further extended the distance, the strength and the speed of which the monkey's brain was capable. Finally, they implemented a bidirectional brain-machine-brain interface within the same individual brain.

Nicolelis grounds his experiments into a theory of the brain that views the neural group (a population of interconnected neurons) as the elementary unit of thought rather than the single neuron. A stimulus has an effect on a neural group: a population of "tuned" neurons is the unit that detects, stores and responds to information. This is a view that he contrasts with the reductionist approach to dig deeper and deeper into smaller and smaller constituents to explain the functioning of the whole. The individual neuron may participate in more than one neural group, and its spiking may have different "meanings" in the various groups. And, viceversa, a particular outcome may be generated by a variety of different neural activities. For example, there are multiple ways that a brain can encode a memory.

Neurons can interact even with distant neurons because the substance enveloping them is highly conductive of electrical signals. This facilitates the creation of countless feedforward and feedback connections, and not only among neighboring neurons. Nicolelis therefore is lukewarm concerning the idea that the cortex is divided into functionally specialized areas.

This broad connectionist view also implies that the brain's working cannot be reduced to a computational algorithm: the human brain is noncomputable.

Nicolelis also emphasizes that the brain is not a passive tool waiting to react to stimuli from the environment but an active information-seeker. And there is vast consensus among biologists that this is true for all forms of life. Nicolelis believes that this active process generates a "brain's point of view" about what is going on in the world and in the body. This nonstop process de facto provides a live simulation of the world, and therefore a set of expectations that in turn determine what the brain seeks for and what the brain will "understand". Hence the brain is three machines in one: a tool to maintain homeostasis in the body (basically, to keep us alive); a simulator to continuously update a model of the world; and an agent to continuously seek information in the environment.

Another interesting finding came from the analysis of brain activity, a finding first presented by Atsushi Iriki ("Coding of modified body schema during tool use by macaque postcentral neurons", 1996). Brains seem to incorporate the tools that the body is using not only in their model of the world but in their very model of the body itself. Tools are represented by the brain as seamless extensions of the body. These "extensions" contributing to one's sense of self would then include clothes, kitchenware, bicycles, cars, digital devices... and even family, friends and acquaintances.

(One could also interview these findings the other way around: perhaps it is not that tools are represented as limbs or organs by the brain but that limbs and organs are represented by the brain just like tools: from the viewpoint of a brain they are all "tools" that help the brain survive, explore the environment, manipulate the environment, fuel itself and reproduce).

Nicolelis envisions a brain-centered future in which one's brain would do everything simply by thinking, without having to rely on body movements and not even on language; in which machines will obey orders received wirelessly from the brain of the master; in which people will be able to communicate remotely with each other without any need to pick up a phone or type an email; in which people will be able to experience the surface of a distant planet while comfortably sitting in a living room's armchair; and in which one would be able to access the thoughts of their dead forefathers and relive vividly their most intimate experiences.

The Origins Of Brains

The anthropological record tells the story of a dramatic increase in brain size that happened over a relatively short period of time. Humans became "intelligent" suddenly and very rapidly. Many scientists and philosophers believe that this rapid increase in brain size must have been driven by a self-amplifying process.

The typical self-amplifying processes are positive feedback loops, in which growth is triggered by growth itself, and there is nothing to stop the growth. The state of the system is out of control. (In negative-feedback systems, such as a furnace for heating a house, there is a control mechanism, e.g. the thermostat. that controls growth for the purpose of maintaining a desired state).

A positive feedback would explain the rapid evolution of the brain. The British philosopher Nicholas Humphrey argues that positive feedback may have come from pressure for "social intelligence", the need to communicate and share with other members of our species. The British biologist Ronald Fisher thought that sexual selection was a form of positive feedback, and this thesis was also defended by the US psychologist Geoffrey Miller.

The US neurobiologist Walter Freeman believes that brains have evolved primarily as organs of social cooperation, and originally they started communicating for sexual reproduction. He retraces the story of socialization, from the early formation of pair bonds and tribal groups and identifies music, dance, and sexually based rituals as the means by which meanings in the brains were shared.

Are Colors More Real Than Pain?

We perceive the world in at least two different ways: one is the sensations that come from the senses (seeing, smelling, touching, hearing, tasting) and one is the feelings that somehow are generated in response to things that are happening to our body (pain, pleasure, hunger, hate, fear, etc).

Colors and shapes seem to be direct perceptions of the world out there, whereas pain seems to be the fictitious outcome of a process in our brain.

It turns out that colors and shapes and sounds and so forth are not shared by all living organisms. As a matter of fact, every species has its own "sensations" that are different from other species. Some animals see the world in three dimensions, but some see it in two dimensions. Some see colors, and some do not. Other species may see "things" that we don't see. Their eyes are different and their brains are different. There is no evidence that what we "see" is what is out there (rather than, say, what the frog sees).

Colors and shapes and sounds and so forth are devices to "map" the outside world so that our body can deal with it in an efficient way. Once we build such a "map" of the outside world, we can, for example, move without hitting solid things and grab things that we want to eat.

Emotions are more of the same, but at a more primitive level. They direct our behavior, but they don't require a representation of the outside world. They just tell us "don't do that" or "do that".

While the ability to see and hear and smell seem to be more primitive than the emotions of pain and pleasure, it is likely to be the opposite.

What is more important for survival? To be able to map the world into objects of such a shape and such a color, or to be able to find food and detect danger in a millisecond?

The early multicellular organisms were probably incapable of any significant mapping of the world around them. They were capable, though, of avoiding dangerous places and moving towards more promising places. They were equipped with simple cells (the progenitors of today's nervous system) that could react to the temperature and to the chemical composition of the surroundings and simply move away or move towards them.

If floating in a pond, those cells would have been able to realize that the temperature was reaching a dangerous level (say, because of a nearby lava flow) and therefore cause movement in the opposite direction. This progenitor of "pain" was much more relevant than assigning a shape or a color to the pond and the lava flow.

The ability to map the world through the senses was a later development, one that provided organisms with an even more sophisticated survival strategy.

Ethical Issues in the Age of the Brain
For centuries the signs of life have been the heartbeat and breathing. It was impossible to measure the mental life of a person, easy to measure heartbeat and breathing.

A person was considered legally dead if the heart stops beating. That is a body-based definition.

Since the 1970s scientists have been able to measure brain activity, not just heartbeat and breathing. In 1963 David Kuhl experimented with emission reconstruction tomography (later renamed Single-Photon Emission Computed Tomography or SPECT). In 1972 Godfrey Hounsfield and Allan Cormack invented X-Ray Computed Tomography Scanning or CAT-scanning; and at the same time Raymond Damadian built the world's first Magnetic Resonance Imaging (MRI) machine. The following year Edward Hoffman and Michael Phelps created the first PET (Positron Emission Tomography) scanner, that allow scientists to map brain function. In 1990 Seiji Ogawa used "functional MRI" to measure brain activity based on blood flow.

Now that it is possible to measure brain activity, in the age in which we assign a higher status to the brain than to other organs, a new brain-based definition of "alive" might be required. The term "brain dead" has come into common use to imply that someone's body is still alive but actually the "person" inside that body is dead.

That leads to new ethical issues. For example, a 12-week fetus has a higher degree of brain activity than an injured person in a persistent and irreversible vegetative stage, but the law grants the latter the full rights of a human being whereas it grants pretty much no rights to the former. Whether either one can be said to be a "thinking", sentient being is debatable, but it is certainly inconsistent that we treat as a human being only the one with lower brain activity. Intensive-care units can keep people's bodies alive but do nothing to resurrect "dead" brains (brains whose thought-related activity is non-existent). The fetus, on the other hand, is living in a sort of "intensive care unit" (the mother's womb) but has a functioning brain. Whether that "functioning" can be said to be "thought" or not is a much bigger issue, but it is at least "more" than what the vegetative person has. Roughly the development of brain activity is: the sub-cortical brain appears at five weeks, the cerebral hemispheres differentiate at seven weeks, EEG activity is present at eight weeks, and the cortex forms at about 20 weeks gestation.

The issue was first advanced by the US pediatrician John Goldenring ("The brain-life theory: towards a consistent biological definition of humanness", 1985). Goldering was basing his opinion piece on flawed data, but he posed the correct question nonetheless: if the cessation of brain activity defines death, shouldn't the onset of brain activity define the beginning of life?

Human cognitive life may be viewed as a continuum from the onset of brain life (about eight weeks gestation) until brain death. In between the brain is definitely active. Before and after that continuum we don't have any evidence that the brain is active other than for mechanical functions. John Goldering's rule of thumb was: "Whenever a functioning human brain is present, a human being is alive"; and viceversa.

According to Michael Gazzaniga (in the "The Ethical Brain", 2005), the human fetus at 13 weeks gestation has brain activity comparable to that of a sea slug; but the brain activity of a vegetative patient is even lower than that. And the near-term fetus has a brain that may well have cognitive capacity since it is more complex than the brain of a sea slug and the sea slug does have a primitive cognitive life.

Furthermore, within that continuum one may encounter cases in which brain activity has fallen below what we consider "human" or never reached that level. If a person's mental life declines to the point that his brain activity is similar to the brain activity of a dog or a rat, should that person still have the full "human rights" or only the rights granted to dog and rats?

Brains are for Traveling and Chatting

What are brains for? Why a rock or a plant can be what it is without a brain, while an animal cannot exist without a brain? What is the unique feature that a brain enables?

One property of living beings is striking. Plants, which do not move, do not have brains. They too grow and they too need to coordinate their growth, but they don't seem to need a brain to do so. Mammals and birds appear to have the most sophisticated ability to move. Mammals and birds also have the most sophisticated brains. The brains of snakes, frogs and fish appear to be simpler, and it turns out that these creatures do not move in as creative a way as mammals and birds. Invertebrates have even simpler brains and their movements are even more basic. Birds and mammals can move very long distances, over huge territories, dealing with a broad spectrum of ecological changes, even crossing oceans and continents, and they can move in a virtually endless variety of ways. Other animals seem to be more limited both in distance and in the degrees of freedom of their movements.

Another property of living beings stands out.

The biggest brain (about 10 kg) belongs to the sperm whale. The record for brain size compared with body mass belongs to the squirrel monkey (5% of the body weight, versus 2% for humans). Some birds too have a higher "brain percentage" than ours (the sparrow is a close second to the squirrel monkey). Two rules seem to tell something about the reasons for larger brains: 1. Bodies of warm-blooded animals consume ten times more energy so brains can be ten times bigger. 2. Species that live in large social groups have the largest brains (the squirrel monkey lives in bands of hundreds of individuals).

That said, let us not forget that the longest living beings on this planet have no brain: trees and bacteria.

Beyond the Brain

How the brain came to be is a long and convoluted story. The brain is made of several regions that evolved at different times and probably for different purposes, and eventually got locked together in the same organ. It is likely that these different circuits had to adapt to each other and to the other organs of the body. Ultimately, the brain had to make sense in the context of itself and of the whole body. What didn't fit in the overall picture was probably fixed by natural selection.

The brain is a Darwinian system, within which the concepts of competition and self-organization are much more important than the ones of design or organization. The brain is a battlefield. Of all possible million configurations at each point in time one is chosen that best fits our experience. Far from being a well-determined logical system, the brain is a chaotic system of trial and error. It is all the more amazing that I can say "i", because my "i", if it is due to my brain, changes all the time, and i have no control over the way my "i" changes: it is experience, not my free will, that selects which connections will get stronger and which ones will die out. I have no more control on the evolution of my thoughts than i have on the evolution of my species.

Over the 19th century we learned to admit our fundamental inability to affect the evolution of our species, and of life in general. A far more powerful force, natural selection, takes care of that. We are simply pawns, created more or less by accident, and doomed to be eventually replaced by other pawns in this eccentric game of life.

Over the 20th century we began to realize that we are also powerless to affect the evolution of our brain, of our own selves. We, at the level of the individual, seem to be in the hands of far more powerful forces that mold our brains regardless of what we would like to be.

Ultimately, our brain is not "ours".

Consciousness is the mysterious entity that somehow enters the picture through the very same brain that we discovered is outside our sphere of influence but that at the same time somehow, we believe, grants us a degree of freedom in thinking, feeling and, ultimately, being what we are.

Further Reading
Baum Eric: WHAT IS THOUGHT? (MIT Press, 2004)
Brown, Jason: THE LIFE OF THE MIND (Lawrence Erlbaum, 1988)
Buck Ross: THE COMMUNICATION OF EMOTION (Guilford Press, 1984)
Changeux, Jean-Pierre: L'HOMME NEURONAL (1983)
Changeux, Jean-Pierre: ORIGINS OF THE HUMAN BRAIN (Oxford University Press, 1995)
Churchland, Paul: ENGINE OF REASON (MIT Press, 1995)
Cytowic, Richard: SYNESTHESIA (1989)
Damasio, Antonio: DESCARTES' ERROR (G.P. Putnam's Sons, 1995)
Delgado, Jose: PHYSICAL CONTROL OF THE MIND (1969)
Edelman, Gerald: NEURAL DARWINISM (Basic, 1987)
Eichenbaum, Howard: COGNITIVE NEUROSCIENCE OF MEMORY (Oxford Univ Press, 2002)

Finger, Stanley: MINDS BEHIND THE BRAIN (Oxford Univ Press, 2004)

Foster Russell & Kreitzman Leon: RHYTHMS OF LIFE - THE BIOLOGICAL CLOCKS THAT CONTROL THE DAILY LIVES OF EVERY LIVING THING (Yale Univ Press, 205)

Freeman, Walter: MASS ACTION IN THE NERVOUS SYSTEM (Academic Press, 1975)

Freeman, Walter: SOCIETIES OF BRAINS (Erlbaum, 1995)

Freeman, Walter: NEURODYNAMICS (Springer, 2000)

Gazzaniga, Michael: THE ETHICAL BRAIN (Univ of Chicago Press, 2005)

Gazzaniga, Michael & LeDoux Joseph: INTEGRATED MIND (Plenum Press, 1978)

Gisolfi, Carl & Mora Francisco: THE HOT BRAIN (MIT Press, 2000)

Goertzel Ben: THE STRUCTURE OF INTELLIGENCE (1993)

Goertzel, Ben: THE EVOLVING MIND (Gordon & Breach, 1993)

Goldstein, Kurt: THE ORGANISM: A HOLISTIC APPROACH TO BIOLOGY (USA Book, 1939)

Hebb, Donald: THE ORGANIZATION OF BEHAVIOR (1949)

Ivry, Richard & Robertson, Lynn: THE TWO SIDES OF PERCEPTION (MIT Press, 1998)

Hebb, Donald: THE ORGANIZATION OF BEHAVIOR (John Wiley, 1949)

Hull, Clark: PRINCIPLES OF BEHAVIOR (Appleton-Century-Crofts, 1943)

James, William: THE PRINCIPLES OF PSYCHOLOGY (1890)

Jerne, Niels: THE NATURAL-SELECTION THEORY OF ANTIBODY (1955)

Joseph, Rhawn: NAKED NEURON (Plenum, 1993)

Koehler, Wolfgang: INTELLIGENSPRUEFUNGEN AM MENSCHENAFFEN (1925)

Lashley, Karl: BRAIN MECHANISMS AND INTELLIGENCE (Dover, 1963)

Lavine, Robert: NEUROPHYSIOLOGY (Collamore, 1983)

Lieberman, Philip: UNIQUELY HUMAN (Harvard Univ Press, 1992)

Livanov, Mikhail: SPATIAL ORGANIZATION OF CEREBRAL PROCESSES (1972)

Llinas, Rodolfo & Churchland Patricia: THE MIND-BRAIN CONTINUUM (MIT Press, 1996)

MacLean, Paul: THE TRIUNE BRAIN IN EVOLUTION (Plenum Press, 1990)

Miller, Geoffrey: THE MATING MIND (Doubleday, 2000)

Nicolelis, Miguel: BEYOND BOUNDARIES (Henry Holt, 2011)
Purves, Dale: NEURAL ACTIVITY AND THE GROWTH OF THE BRAIN (Cambridge Univ Press, 1994)
Ramachandran, Vilayanur & Blakeslee, Sandra: PHANTOMS IN THE BRAIN (Morrow, 1998)
Ramachandran, Vilayanur: THE TELL-TALE BRAIN (Norton, 2011)
Rizzolati, Giacomo: MIRRORS IN THE BRAIN (Oxford Univ Press, 2008)
Skinner, Burrhus: BEHAVIOR OF ORGANISMS (1938)
Swanson, Larry: BRAIN ARCHITECTURE (Oxford Univ Press, 2002)
Thorndike Edward: ANIMAL INTELLIGENCE (1911)
Underwood Geoffrey: OXFORD GUIDE TO THE MIND (Oxford Univ Press, 2000)
Valiant, Leslie: CIRCUITS OF THE MIND (Oxford University Press, 1994)
Ramachandran, Vilayanur & Blakeslee, Sandra: PHANTOMS IN THE BRAIN (Morrow, 1998)
Young, John: A MODEL OF THE BRAIN (Clarendon Press, 1964)
Zeki, Semir: A VISION OF THE BRAIN (Blackwell, 1993)

Memory: The Mind's Growth

Memories are Made of This

The mind's cognitive faculties depend to a great extent on memory. If we could not learn and remember at all, our cognitive life would be virtually non-existent.

If we could not remember where we live, where our office is, how to lace our shoes, how to drive a car, how to speak, and so forth, we would be mere objects devoid of real life. In fact, we would probably last only a few minutes. The more complex the organism, the more essential memory is to its survival because so much is required to keep the organism alive.

Even the worst cases of amnesia do not completely erase memory. A patient who suffers complete amnesia does not remember anything from some date on in the past, but still remembers a lot of vital facts about living in the world.

Evolutionarily speaking, memory provided a considerable advantage to creatures capable of remembering where water was or where predators lived. The more refined one's memory, the easier to navigate the environment, to survive in it and to find food in it.

That animal memory is not just like a computer memory is a fact due precisely to the tasks that were required by it, tasks that rarely require perfection but do require speed and capacity. Learning to bike and memorizing the emergency phone number are both important tasks, but they are usually achieved in rather different ways. Thus it is not even correct to speak of "memory" as if it were just one task. It is probably more appropriate to speak of "memories".

Memory is more than storage. Memory is also recognition. We are capable of recognizing a tree as a tree even if we have never seen that specific tree before. No two trees are alike. And even a specific tree never appears the same to us, as the perspective, the wind, the lighting can all dramatically change its appearance. In order to recognize a tree as a tree, and as a specific tree, we use our "memory". Whenever we see something, we ransack our memory looking for "similar" information. Without memory we would not see trees, but only patches of brown and green.

The process of thinking depends on the process of categorizing: the mind deals with concepts, and concepts exist because memory is capable of organizing experience into concepts. Our mind, ultimately, looks like a processor of concepts. The mind's functioning is driven by memory, which is capable of organizing knowledge into concepts. So much so that, inevitably, a theory of memory becomes a theory of concepts, and a theory of concepts becomes a theory of thought.

Cognition revolves around memory. All cognitive faculties use memory and would not be possible without memory. They are, in fact, but side effects of the process of remembering. There is a fundamental unity of cognition, organized around the ability to categorize, to create concepts out of experience.

Memory's task is easily summarized: to remember past experience. But, unlike the memory of a computer, which can remember exactly its past experience, human memory never remembers exactly.

The most peculiar feature of our memory is, perhaps, the fact that it is so bad at remembering. Our memory does not only forget most of the things that happen, but, even when it remembers, it does a lousy job of remembering.

Memory of something is almost always approximate. Many details are forgotten right away. If we want to remember a poem by heart, we have to repeat it to ourselves countless times. And sometimes memory is also very slow: sometimes it takes a long time to retrieve a detail of a scene, sometimes it will take days before the name of a person comes back to mind. Rather than accessing memories by calendar day or person's name, we seem to access them by associations, which is a much more complicated way to navigate in the past.

It is hard to think of something without thinking also of something else. It is hard to focus on a concept and not think of related concepts. And the related concepts that come to mind when we focus on a concept are usually things we care about, not abstract ideas. If we focus on "tree", we may also remember a particular hike in the mountains or an event that occurred by a tree. We build categories, we relate categories among them, we associate specific episodes with categories.

For an entity that is supposed to be just a storage device, anomalies abound. For example, we cannot count very easily. Do you know how your home looks like? Of course. How many windows does it have? You have looked at your home thousands of times, but you cannot say for sure how many windows it has. If you see a flock of birds in the sky, you can tell the shape, the direction, the approximate speed... but not how many birds are in the flock, even if there are only six or seven. Another weird feature of our memory is that it is not very good at remembering the temporal order of events: we have trouble remembering if something occurred before or after something else. On the other hand, our memory is good at ordering objects in space and at counting events in time.

Human memory is a bizarre device that differs in a fundamental way from the memory of machines: a camera or a computer can replicate a scene in every minute detail, whereas our memory was just not designed to do that.

What was our memory designed to do?

The Reconstructive Memory

A startling feature of our memory is that it does not remember things the way we perceived them. Something happens between the time we see or hear a scene and the time that the scene gets stored in memory.

I can tell you the plot of a novel even if i cannot tell you a single sentence that was in the novel. If I tell you the plot twice, i will use different words. It would be almost impossible to use the same words. Nobody can remember all the sentences of a book, but everybody can remember what the "story" is.

Compare with a computer. A computer can memorize the book page by page, word by word. Our memory does not memorize that way. It is not capable of memorizing a book page by page, word by word. On the other hand, it is capable of so many other things that a computer is not capable of. For example, we can recognize a plot, told by somebody else, as the plot of the same novel. That person's version of the plot and our version of the plot probably do not share a single sentence. Nonetheless, we can recognize that they are the same story. No computer can do that (yet), no matter how big its memory is. Size is obviously not the solution.

In the 1930s the British psychologist Frederic Bartlett developed one of the earliest models of memory. Bartlett studied how memory "reconstructs" the essence of a scene. We can easily relate the plot of a movie, and even discuss the main characters, analyze the cinematography, and so forth, but we cannot cite verbatim a single line of the movie. We stored enough information about the movie that we can tell what it was about and perform all sorts of reasoning about it, but we cannot simply quote what a character said at one point or another.

What Bartlett discovered is that events are not stored faithfully in memory: they are somehow summarized into a different form, a "schema". Individuals do not passively record stories verbatim, but rather actively code them in terms of schemas, and then can recount the stories by retranslating the schemas into words.

Each new memory is categorized in a schema which depends on the already existing schemas. In practice, only what is strictly necessary is added. When a memory must be retrieved, the corresponding schema provides the instructions to reconstruct it. That is the reason why recognizing an object is much easier in its typical context than in an unusual context.

The advantage of the "reconstructive" memory is that it can fit a lot of information in a relatively narrow space. Any memory that tried to store all the scenes, text and sound of a movie would require an immense amount of space. But our memory stores only what is indispensable for reconstructing the plot and other essential features of the movie, thereby losing lots of details but at the same time saving a lot of space.

Reconstructing And Making Sense

The mechanism of "reconstructing" the memory of an event is quite complex. There is more involved than a simple "retrieval" of encoded information.

In 1904 German biologist Richard Semon had already speculated that memory was as much about retrieval as about storage. He introduced the concept of "engram": the unit of memory, or, better, the pattern used to encode it (the "memory trace"). He then introduced another concept: the "ecphoric stimulus": the cue that helps retrieve a specific memory. He noticed that the likelihood of finding a memory depends also on the cue that is used to retrieve it (the pattern used to decode it). We are often forced to remember something simply because we encountered a word or saw something that "reminds" us of something else. It was merely a fleeing moment, but enough to bring back the memory of something or somebody. Semon realized the power of cues: a cue is only a fraction of the engram, but it is enough to retrieve the whole engram.

The fact that memory is not a linear recording of sensory input reveals that something helps memory make sense of the past. When memory reconstructs an event, it must have a way to do so in a "meaningful" way. The memory of an event is not just a disordered set of memories more or less related to that event. It is one flowing sequence of memories that follow one from the other. Sometimes you cannot finish relating the plot of a novel because you "forgot" a key part of it: the truth is that you forgot all of it and you were reconstructing it, and, while reconstructing it, you realized that something was missing. You cannot reconstruct the plot because you have an inner sense of what reality must be like. You may know how the novel ends and how it goes up to a point, and then you realize that something is necessary in order to join that point to the ending. Your reconstructive memory knows that something is missing in the reconstruction because the reconstruction does not yet "make sense".

The fascination of movies or novels is that you have to put together reality until it makes sense again. You have to find the missing elements so that the story gets "explained". Our brain has a sense of what makes sense and what does not.

Cognitive Maps

The US gestalt psychologist Edward Tolman ("Cognitive Maps in Rats and Men", 1948) can be credited with coining the concept of a "cognitive map": a rat knows how to navigate a maze because it maintains a cognitive map that covers much more than the rat has ever experienced directly. Tolman proved that rats build a cognitive map of an environment even without any reward, simply because they "were there".

A cognitive map is a mental representation of the world in which we live.

Cognitive maps both represent and participate in the creation of our experience of the world. A cognitive map is created and continuously improved through the individual's experience and by interaction with other cognitive maps. At the same time, the cognitive map "is" the world, insofar as the individual is concerned. The map is how the world appears to be to the individual. The individual only knows her or his map of the world. This map thus works as an anticipatory schema, that determines what we expect to see and, ultimate, what we indeed see.

The Partitioning of Memory

The works of George Miller ("The Magical Number Seven, Plus or Minus Two", 1956), Donald Broadbent (1957), Allen Newell (1958) and Noam Chomsky (1957), that all came out in the second half of the 1950s, established a new paradigm in Psychology, broadly referred to as "cognitivism", that ended the supremacy of behaviorism. Where behaviorism was only interested in the relationship between input (stimuli) and output (behavior), cognitivism focused on the processing that occurs between the input and the output

Herbert Simon and Alan Newell argued that the human mind is a symbolic processor and Noam Chomsky advanced a theory of language based on modules inside the mind which are capable of symbolic processing.

The standard model that became popular in the late 1950s, due particularly to the work of the British psychologist Donald Broadbent and the US psychologist George Miller, was based on the existence of two types of memory: a "short-term memory", limited to few pieces of information, capable of retrieving them very quickly and subject to decaying also very quickly; and a "long-term memory", capable of large storage and much slower in both retrieving and decaying. Items of the short-term memory move to the long-term memory after they have been "rehearsed" long enough.

The idea was already implicit in William James' writings (he called them "primary" and "secondary" memory), and in Theodore Ribot's 1882 experiments on amnesia (the loss of memory is inversely proportional to the time elapsed between the event and the injury), but Broadbent also hypothesized that short-term memory may just be a set of pointers to blocks of information located in long-term memory.

Broadbent also stated the principle of "limited capacity" to explain how the brain can focus on one specific object out of the thousands perceived by the retina at the same time. The selective character of attention is due to the limited capacity of processing by the brain. In other words, the brain can only be conscious of so many events at the same time. What actually gets the attention is complicated to establish, because Broadbent found out that attention originates from a multitude of attentional functions in different subsystems of the brain.

Broadbent's model of memory (also known as the "filter theory") reflected at least two well-known features of memory: information about stimuli is temporarily retained but it will fade unless attention is turned quickly to it; the unattended information is "filtered out" without being analyzed. He drew a distinction between a sensory store of virtually unlimited capacity and a "categorical" short-term store of limited capacity. The latter is the way that a limited-capacity system such as human memory can cope with the overwhelming amount of information available in the world.

At the same time, George Miller's experiments proved that our short-term memory can hold only up to seven "chunks" of information and therefore provided an order of magnitude for it. It wasn't clear, though, what was the "size" of a chunk: is the entire car a chunk of information, or is each wheel a chunk, or...? In Broadbent's model, a chunk is a pointer to something that already exists. Therefore a chunk can be even very "big", as long as it is already in memory. Its "size" is not important (in short-term memory, it is only a pointer). This is consistent with experiments in which short-term memory proves to be capable of holding familiar images, but not of images never seen before.

The British psychologist Alan Baddeley ("Working Memory", 1974) showed that a unitary short-term memory does not account for memory disorders and replaced short-term memory with a "working memory" that has basically three components: a short-term memory for verbal information, a short-term memory for visual information, and a control system.

In fact, neural regions in the prefrontal cortex (the newest part of the brain, from an evolutionary standpoint) can draw data from other regions of the brain and hold them for as long as needed. The prefrontal cortex is unique in having a huge number of connections with the sensory system and with lower brain centers. The prefrontal cortex could be the locus of a "working memory", in which decisions, planning and behavior take place.

Types of Memory

Experiments performed in the 1970s by the Canadian psychologist Endel Tulving and his associate Daniel Schacter proved that "intension" (such as concepts) and "extension" (such as episodes) are dealt with by two different types of memory.

"Episodic" memory contains specific episodes of the history of the individual, while semantic memory contains general knowledge (both concepts and facts) applicable to different situations. Episodic memory, which receives and stores information about temporally-dated episodes and spatiotemporal relations among them, is a faithful record of a person's experience.

"Semantic" memory, instead, is organized knowledge about the world. Tulving believes these memory systems are physically distinct because their behavior is significantly different. In episodic memory, for example, the recall of a piece of information depends on the conditions ("cues") under which that piece of information has been learned (an explicit or implicit reference to it).

There are at least two more aspects of memory that fall neither into the intension or extension.

Procedural memory allows us to learn new skills and acquire habits. William James had been particularly interested in this kind of memory, having realized how important "habits" are to determine our behavior. He reduced habits to a sequence of "reflexes", i.e. stimulus-response events. Basically, each stimulus-response pattern, once learned, becomes the building block for more complex patterns which are our "habits", each of which is in turn a building block to create more complex "habits". The French philosopher Henri Bergson explicitly separated the memory of habits from the memory of events.

As the French philosopher Maine de Biran had already observed two centuries earlier ("The Influence of Habit on the Faculty of Thinking", 1804), habits rely on an "implicit" memory.

Implicit memory is "unconscious" memory, memory without awareness: unlike other types of memories, retrieval cues do not bring about a recollection of them. Implicit memories are weakly encoded memories which can nonetheless affect conscious thought and behavior. Implicit memories are not lost: they just cannot be retrieved. Amnesia is the standard condition of human memory: most of what happens is not recorded in a form that can be retrieved. In the first years, because of incompletely developed brain structures, most memories are lost or warped. Nonetheless, memories of childhood are preserved without awareness of remembering. Implicit memory is the one activated in "priming" events, or in the identification of words and objects.

That makes a grand total of four different types of memory: procedural, semantic, episodic and implicit.

Tulving also devised a scheme by which memory can associate a new perception or thought to an old memory: the remembering of events always depends on the interaction (or compatibility) between encoding and retrieval conditions.

It indeed appears that the brain accomodates several different memory systems, each of them involving the cortex but each characterized by different "pathways" leading from the cortex to other areas of the brain. Studies on amnesia (particularly by Neal Cohen in 1980) show that there are at least two separate memory systems: "declarative" memory (the memory that one can consciously remember, which is forgotten in an amnesia) and "procedural" memory (the skills and procedures which are usually not forgotten, as people with amnesia can still perform most actions they have learned throughout their lives). It appears that the hippocampus is the key to declarative memory, or at least the key to linking together declarative memories. Procedural memory is, instead, realized by circuits that involve the motor areas of the cortex and two loops that spread through the striatum and the cerebellum: acquiring skills is, indeed, a complex phenomenon.

"Emotional" memory, on the other hand, seems to depend on the working of the amygdala, i.e. on yet another separate memory system. These three memory systems are physically connected to the cortex along different pathways, which means that they can work in parallel.

The Rememberer

Tulving summarized the relationship between remembered and rememberer in the "encoding specificity principle": remembering depends on the affinity between encoding and decoding. Memories are encoded in a way that depends on the circumstances when the event originally happened. The likelihood of recalling a memory (of decoding it) depends on recreating those circumstances, on reinstating the same psychological state. In other words, the way we feel about an event plays an important role in the way that event can later be recalled. For example, the feeling that I feel when I read a sentence is going to be important for later recalling that sentence. That feeling has become part of the episode, as it is encoded in my memory.

Tulving's episodic memory packages different aspects of an event to give it the "autobiographical" feeling that makes it more than just a retrieval of information, it makes it a memory of something that happened in our life. In other words, an essential part of an episodic memory is the "rememberer". The rememberer does more than retrieve information about a past event: the rememberer experiences that event again. In fact, the episodic memory is more about the feeling of being there than about the event in itself: the feeling of the event is generally recalled in more accurate terms than the details of the event. In fact, it is easier to remember something that happened a long time ago but had a strong emotional impact on us than something that happened just minutes ago. I do not remember what I had for lunch two days ago, but I do remember episodes of my childhood that happened several decades ago (if I focus, I can even feel what I felt then). Ultimately, episodic memory is about the rememberer, not the remembered.

Confusing the Self

The US psychologist Daniel Schacter believes that "memory" is actually a set of different kinds of memory, each specialized in a different kind of task and implemented by a different brain circuit.

The memory system that we usually refer to is the "explicit memory". Implicit memory operates outside our awareness. Semantic memory handles conceptual knowledge. Procedural memory is used to learn skills and acquire habits. Episodic memory recalls episodes. Schacter believes that each is implemented in a different memory system, each having different implications for the rest of the brain.

Rephrasing Endel Tulving's ideas on encoding and decoding, Schacter makes a distinction between "field memory" (you are in it) and "observer memory" (you are not in it). We tend to recall older events as field memories, and more recent ones as observer memories. But this is not a given: it really depends on what we focus on. If asked to focus on the details of an event, we recall them as observer memories. If asked to focus on our feelings during the event, we tend to switch to a field memory. Thus the "perspective" is not part of the stored information: the perspective is manufactured at the time of recall.

Schacter makes the point that memory does not store events as they happened (a memory of a scene is not a snapshot of that scene), nor does it store the event per se: it stores our experience of them. That experience is more than just a recording of what happened. It contains information on how to operate in a similar situation. Memory of the past determines how we act in the future.

Subjective experience is crucial to the process of remembering: it literally shapes what we remember.

Memories are somehow encoded. Thus they have to be decoded whenever needed. But the decoding does not seem to be deterministic. And the encoding is likely to be very subjective (two people encode the same event in two different ways, and the very same person may encode the same event in two different ways if it occurs twice). Schacter thinks that we remember what we analyzed, and that the process of encoding an event is simply a by-product of our analysis of that event.

The retrieval is no less complicated than the original encoding. Schacter (following his fellow psychologist Morris Moscovitch) distinguishes at least two kinds. "Associative retrieval" is the kind of remembering that we can't help: we are reminded of something, whether we want to or not. "Strategic retrieval" is the kind of remembering that we do when we try very hard to remember something (a name, for example). Moscovitch believes that the former is mainly carried out by the hippocampus, while the latter is probably performed in the cortex: basically, these are different processes that take place in different places.

Schacter believes that the cue is also part of the "remembered": the cue combines with the engram (the memory trace) to yield a subjective recollection. Basically, depending on which cue triggers the "remembering" of an event, we remember that event in a slightly different light. "A memory is an emergent property of the cue and the engram".

This is also why we forget: new experiences blur previous engrams. Cues that used to be very effective become less and less effective.

At the same time, we consolidate memories, and they become more fault-tolerant. This happens mainly through "reuse" of them: as we use them, we rehearse the connection between the medial temporal region (the "index" of features that are stored in the cortex, what Antonio Damasio calls "convergence zone") and the storage areas in the cortex (that contain those features). Basically, when we use and reuse a memory, the medial temporal region rehearses how to reconstruct that memory. The hippocampus also contributes to consolidating memories, as per Jonathan Winson's theory of dreams.

Thus different aspects of a memory are stored in different regions of the brain, and then connected by a special memory system, which is therefore responsible for the "autobiography" that each of us has of her/himself.

Schacter believes that there are three levels of autobiographical knowledge: lifetime periods (e.g., childhood), general events (e.g., a holiday) and events (e.g., the first kiss). Each remembered episode is actually a mixture of these three levels.

Schacter makes the point that, because the encoding process is subjective, there is a limit to how accurate the decoding can be. Just like rehearsal can increase the likelihood of retrieving a memory, it can also increase its inaccuracy. We distort our memories of ourselves. We construct our own autobiography, which is only loosely based on what truly happened.

Categories

Arguably the most important function of memory is categorization. The rings of a tree or the scratches on a stone can be said to "remember" the past, but human memory can do more: it is capable of using the literal past to build abstractions that are useful to predict the future. It is able to build generalizations. Actually, categorization is the principal way that humans have of making sense of their world. For example, if we analyze the grammar of our language, the basic mechanisms of meaning-bearing are processes of categorization.

One can even wonder whether all living beings, or at least many of them, need to achieve some level of categorization in order to deal with the world.

The German linguist Eric Lenneberg argued that all animals organize the sensory world through a process of categorization. They exhibit propensities for responding to categories of stimuli, not to single specific stimuli. In humans this process of categorization becomes "naming", ie, the ability to assign a name to a category. But even in humans the process of categorization is still a process whose function is to enable "similar" response to "different" stimuli. For example, we "sit" on a "chair", regardless of how different this chair is from other chairs on which we sat in the past.

Traditionally, categories were conceived as being closed by clear boundaries and defined by common properties of their members. In the 1950s the US psychologist Jerome Bruner proposed that one could conceive of categories as sets of features: a category is defined by the set of features that are individually necessary and jointly sufficient for an object to belong to it. In order words, one can write down the rules that specify what is necessary and sufficient for a member to belong to a category.

This seems to be the case for nominal types (the one invented by us, such as "mother" or "triangle"), but not necessarily for natural types. As the great Austrian philosopher Ludwig Wittgenstein pointed out, a category like "sport" does not fit the classical idea (both cards and chess and football are sports, but they have very little in common). A dog that does not bark or a dog with three legs or a vegetarian dog would probably still be considered a dog, even if it violates the set of features we usually associate with the concept of a dog. What unites a category is "family resemblance", plus sets of positive and negative examples. Its boundaries are not important: they can be extended at any time.

Bruner was among the first to realize that most cognitive processes are nothing but classification processes in disguise. Cognitive activity ("thinking") depends on placing an event or situation in the appropriate category. A category is basically a set of events that can be treated the same way by the cognitive organism. Bruner also realized that categories are not "discovered" but "invented". They do not exist in the environment: they are construed by the human mind. Thus the inferences that matter are really the one that helps create a new category based on some events, and the inference that helps classify an event relative to the existing categories. A concept is a network of such inferences that allow us to infer an event's category based on some observed attributes of the event, and then to infer the unobserved attributes of that event. His "functionalist" definition of an object, for example, is the network of inferences about it that one is capable of employing after an act of categorization. There exist different kinds of concepts that employ and trigger different kinds of inferences A common kind of inference is the one that all members of a category share some common attributes, but this is only one of the possible inferences to define a category.

Prototype Theory

The traditional view that categories are defined by common properties of their members was quickly replaced by Eleanor Rosch's theory of prototypes. After all, the best way to teach a concept is to show an example of it.

The US psychologist Eleanor Rosch noted that some members of a category seem to be better examples of the category than others. Not all members are alike, even if they all share the same features of the category. This means that the features by themselves are not enough to determine the category. It also implies that there must exist a "best" example of the category, what she called the "prototype" of the category.

In the 1970s she founded her early theory on two basic principles of categorization: 1. The task of category systems is to provide maximum information with the least cognitive effort; and 2. The perceived world comes as structured information. In other words, we do categorization because it helps save a lot of space in our memory and because the world lends itself to categorization. Concepts promote a cognitive economy by partitioning the world into classes, and therefore allowing the mind to substantially reduce the amount of information to be remembered and processed.

In Rosch's theory of prototypes, a concept is represented through a prototype that expresses its most significant properties. Membership of an individual in a category is then determined by the perceived distance of resemblance of the individual to the prototype of the category.

Next, Rosch proposed that thought in general must be organized around a privileged level of categorization. In the 1950s the US psychologist

Roger Brown ("Words and Things", 1958) had noted that children tend to learn concepts at a level which is not the most general and not the most specific: say, "chair", rather than "furniture" or "armchair". And in the 1960s the US anthropologist Brent Berlin through his studies on colors and on plant-naming and animal-naming ("Covert Categories and Folk Taxonomies", 1968) had reached a similar conclusion that applies to categories used by adults. The point was that we can name objects in many different ways: a cat is also a feline, a mammal, an animal, and it is also a specific variety of cat. However, we normally call it "a cat". The level at which we "naturally" name objects is the level of what Brown termed "distinctive action". The actions we perform on flowers are pretty much all the same, and certainly different from the actions that we perform on a cat (e.g., one we smell and one we pat). But the actions we perform on two different varieties of cats or two different types of flowers are the same (we pat both the same way, we smell both the same way). Our basic actions tell us that a cat is a cat and a flower is a flower, but they cannot tell us that a rose is not a lily. "Cat" and "flower" represent a "natural" level of categorization.

Berlin had found that people categorize plants at the same "basic level" anywhere in the world (which roughly corresponds to the genus in biology). It is a level at which only shape, substance and pattern of change are involved, while no technical details are required.

Rosch extended his ideas to artifacts and she found that we also classify artifacts at a "basic level" where technical details are not essential. We first create categories of "chair" and "car", and only later we specialize and generalize those categories (to "armchair", "furniture", "sport car", etc). At the basic level we can form a mental image of the category. We can form a mental image of "chair", but not of "furniture". We can form a mental image of "car", but not of "vehicle". We have a motor program for interacting with "chair", but not with "furniture". We have a motor program for interacting with "car", but not with "vehicle". Categorization initially occurs based on our interaction with the object. Meaning is in the interaction between the body and the world.

Rosch postulated a level of abstraction at which the most basic category cuts are made (i.e., where "cue validity" is maximized), which she called the "basic" level. Categories are not merely organized in a hierarchy, from more specific to more general. There is one level of the hierarchy that is somewhat privileged when it comes to perception of form, movement of body parts, organization of knowledge, etc. "Chair" and "car" are examples of basic categories. We can form a mental picture of them. We have a motor program for dealing with them. They are the first ones learned by children. The category of "furniture", for example, is different:

I cannot visualize it, I do not have a motor program to deal with it, and it takes some time for a child to learn it.

Generalization tends to proceed upwards from this level, and specialization proceeds downward from this level. Superordinate categories are more abstract and more comprehensive. Subordinate categories are less abstract and less comprehensive. The most fundamental perception and description of the world occurs at the level of basic (or natural) categories.

Rosch also realized that categories occur in systems, not alone, and they depend on the existence of contrasting categories within the same system. Each contrasting category limits a category (e.g., if a category for birds did not exist, the category for mammals would probably be bigger). At the basic level, categories are maximally distinct, i.e. they maximize perceived similarity among category members and minimize perceived similarities across contrasting categories. Technically, one can use the notion of "cue validity": the conditional probability that an object falls in a particular category given a specific feature. Category cue validity is the sum of all the individual cue-validities of the features associated with a category. The highest cue validity occurs at the basic level. The lowest cue-validities are those for super-ordinate categories.

Fuzzy Concepts

Later, Rosch recognized that categories are not mutually exclusive (an object can belong to more than one category to different degrees), i.e. that they are fundamentally ambiguous. This led to the use of fuzzy logic in studying categorization.

For example, the US linguist George Lakoff borrowed ideas from Wittgenstein's family-resemblance theory, Rosch's prototype theory and Lotfi Zadeh's theory of fuzzy quantities for his theory of "cognitive models".

Lakoff started off by demolishing the traditional view of categories: that categories are defined by common features of their members; that thought is the disembodied manipulation of abstract symbols; that concepts are internal representations of external reality; that symbols have meaning by virtue of their correspondence to real objects.

Lakoff showed that categories depend on two more factors: the bodily experience of the "categorizer" and the "imaginative processes" (metaphor, metonymy, mental imagery) of the categorizer.

Lakoff's theory is based on the assumption of "embodiment of mind": there is no green in the world, but green has to do with the relationship between my body (my eye, my retina, my brain) and the world. Meaning cannot be in the world because things are not in the world: they are in the relationship between us and the world.

His close associate, the US philosopher Mark Johnson, had shown that experience is structured in a meaningful way prior to any concepts: some schemas are inherently meaningful to people by virtue of their bodily experience (e.g., the "container" schema, the "part-whole" schema, the "link" schema, the "center-periphery" schema). We "know" these schemas even before we acquire the related concepts because such "kinesthetic" schemas come with a basic logic that is used to directly "understand" them.

Thus Lakoff argued that thought makes use of symbolic structures which are meaningful to begin with (they are directly understood in terms of our physical experience): "basic-level" concepts (which are meaningful because they reflect our sensorimotor life) and kinesthetic image schemas (which are meaningful because they reflect our spatial life). Other meaningful symbolic structures are built up from these elementary ones through imaginative processes such as metaphor.

As a corollary, everything we use in language, even the smallest unit, has meaning. And it has meaning not because it refers to something, but because it is either related to our bodily experience or because it is built on top of other meaning-bearing elements.

Thought is embodiment of concepts via direct and indirect experience. Concepts grow out of bodily experience and are understood in terms of it. The core of our conceptual system is directly grounded in bodily experience. This explains why Rosch's basic level is what it is: the one that reflects our bodily nature. Meaning is based on experience. With Putnam, "meaning is not in the mind". But, at the same time, thought is imaginative: those concepts that are not directly grounded in bodily experience are created by imaginative processes such as metaphor.

Knowledge is organized into categories by what Lakoff calls "idealized cognitive models". Each model employs four kinds of categorizing processes: "propositional" (which specifies elements, their properties and relations among them in a manner similar to frames); "image-schematic" (which specifies spatial images in a manner similar to Ronald Langacker's image schemas); "metaphoric" (which maps a propositional or image-schematic model in one domain to a model in another domain); and "metonymic" (which maps an element of a model to another element of the same model).

Some models are classical (in that they yield categories that have rigid boundaries and are defined by necessary and sufficient conditions), some models are scalar (they yield categories whose members have only degrees of membership). All models are embodied, i.e. they are linked with bodily experience.

Models build what the French linguist Gilles Fauconnier calls "mental spaces", interconnected domains that consist of elements, roles, strategies and relations between them. Mental spaces allow for alternative views of the world. The mind needs to create multiple cognitive spaces in order to engage in creative thought.

Lakoff argues that the conceptual system of a mind, far from being one gigantic theory of the world, is normally not consistent. We have available in our minds many different ways of making sense of situations. We constantly keep alternative conceptualizations of the world.

The Origin of Categories

The 18th century German philosopher Immanuel Kant held that experience is possible only if we have knowledge, and knowledge evolves from concepts. Some concepts must therefore be native. We must be born with an infrastructure that allows us to learn concepts and to build concepts on top of concepts.

Chomsky proved something similar for language: that human brains are designed to acquire a language, that they contain a "universal grammar" ready to adopt the specific grammar of whatever language we are exposed to. We speak because our brain is meant to speak.

Kant, in a sense, stated the same principle for thinking in general: we think in concepts because we are meant to think in concepts. Our mind creates categories because it is equipped with some native categories and a mechanism to build categories on top of existing categories.

Just like Chomsky said that grammar is innate and universal, so one can claim that some concepts are innate and universal.

Conceptual Holism

Inspired by Willard Quine's holism ("Two Dogmas of Empiricism", 1951), the US psychologist Frank Keil argues that concepts are always related to other concepts. No concept can be understood in isolation from all other concepts. Concepts are not simple sets of features. Concepts embody "systematic sets of causal beliefs" about the world and contain implicit explanations about the world. Concepts are embedded in theories about the world, and they can only be understood in the context of such theories.

In particular, natural kinds (such as "gold") are not defined by a set of features or by a prototype: they derive their concept from the causal structure that underlies them and explains their superficial features. They are defined by a "causal homeostatic system", which tends to stability over time in order to maximize categorizing.

Nominal kinds (e.g., "odd numbers") and artifacts (e.g., "cars") are similarly defined by the theories they are embedded in, although such theories are qualitatively different. There is a continuum between pure nominal kinds and pure natural kinds with increasing "well-definedness" as we move towards natural kinds.

What develops over time is the awareness of the network of causal relations and mechanisms that are responsible for the essential properties of a natural kind. The theory explaining a natural kind gets refined over the years.

The Mind's Growth

The fundamental feature of the mind is that it is not always the same. Just like every other organ in the body, it undergoes growth. It is not only a matter of memory getting "bigger": the "quality" of the thought system changes in a significant way. What we are capable of doing with our minds changes dramatically during the growth of the mind from childhood to adulthood. It is more than just learning about the environment: the mind literally "grows" into something else, capable of new types of actions. The brain, as well as the rest of the body, undergoes a massive change in shape and volume. Somehow this also results in significant new skills.

The balance between "nature" and "nurture" (between "nativism" and "constructivism") is the key to understanding the mind's growth. Humans are born with "instincts", and then "experience" shapes the mind, i.e. nature and nurture coexist and interact.

Constructivism

In the 1930s the Swiss psychologist Jean Piaget introduced an important framework to study the growth of mind.

The biological context of his ideas is that living beings are in constant interaction with their environment, and survival depends on maintaining a state of equilibrium between the organism and the environment. The organism has to regulate its own behavior in order to continuously adapt to the information flow from the environment. At the same time the behavior of the organism shapes the environment, and, of course, the aim of the organism is to shape the environment so as to maximize the chances of maintaining the vital equilibrium.

Cognition, therefore, is but self-regulation.

A dynamic exchange between organism and environment is also the basis of his theory of knowledge, which he labeled "genetic epistemology". The cognitive process (the self-regulation) consists in a loop of assimilation and accomodation.

This process occurs in stages. The development of children's intellect proceeds from simple mental arrangements to progressively more complex ones not by gradual evolution but by sudden rearrangements of mental operations that produce qualitatively new forms of thought.

Cognitive faculties are not fixed at birth but evolve during the lifetime of the individual.

First a child lives a "literal" sensorymotor life, in which knowledge of the world is only due to her actions in it. Slowly, the mind creates "schemas" of behavior in the world. Autonomous, self-regulated functioning of "schemas" lead to "interiorized" action. The child begins to deal with internal symbols and introspection. Then the child learns to perform internal manipulations on symbols that represent real objects, i.e. internal action on top of external action. Finally, the mental life extends to abstract objects, besides real objects. This four-step transition leads from a stage in which the dominant factor is perception, which is irreversible, to a stage in which the dominant factor is thought, which is reversible.

Language appears between the sensorymotor and the symbolic stages, and is but one of the elements of symbolic thought.

As opposed to Jerry Fodor's innate "language of thought", symbolic representation is constructed during the child's development.

The mind's growth is due to the need to maintain a balance between the mind and its knowledge of the world. Rationality is the overall way in which an organism adapts to its environment. Rational action occurs every time the organism needs to solve a problem, i.e. when the organism needs to reach a new form of balance with its environment. Once that balance has been achieved, the organism proceeds by instinct. Rationality will be needed only when the equilibrium is broken again.

In conclusion, Piaget did not recognize a major role for any innate knowledge. He only accepted a set of sensory reflexes and three processes: assimilation, accomodation and equilibration. These processes are very general, not specific to any domain. The same processes are supposed to operate on development of language, reasoning, physics, etc. Piaget's child is a purely sensorymotor device.

Piaget's stance is almost behavioristic, except that he grants the child an inner growth.

From The Social To The Personal
By studying the behavior of chimpanzees, the Russian psychologist Lev Vygotsky reached the conclusion that thought and speech originate from different processes, and then evolve in parallel but independently of each

other. The close correspondence between thought and speech is unique to adult humans.

Children initially behave like chimpanzees: language and thought are unrelated (language is irrational, thought is nonverbal). They learn the names of objects only when told so. At some point the attitude changes: it is the child who becomes curious about the names of things. At that point the child's vocabulary increases dramatically, with much less coaching from adults. The child has learned that objects have names, or, equivalently, that one of the properties of an object is its name. At this point in the development of the child, thought and speech merge.

Vygotsky redrew Piaget's theory of egocentric speech (the kind of speech that ignores the rest of the world) in pre-school children. One generally becomes aware of her/his actions when they are interrupted. Speech is an expression of the process of becoming aware of one's actions. The egocentric child is no exception: its egocentric speech is the sign of a process of becoming aware after something disrupted the action underway. In other words, the child is thinking aloud. A few years later this process has become silent: when the child needs to find a solution to a problem, the "thinking" is no longer aloud, has become an inner conversation. When egocentric speech disappears, it still exists, but has moved inside. The reason it is no longer "vocal" is because it does not serve a social function anymore (it does not need to be heard by others).

According to Piaget, social speech follows egocentric speech, but Vygotsky believes that speech is originally social in nature, and egocentric speech is a specialization of it used when the child has to reflect. Egocentric speech is an evolution of social speech that eventually becomes silent thought. Cognitive faculties are internalized versions of social processes.

The unit of verbal thought is word meaning, which also represents the fusion of thought and speech. Adults and children use the same word to refer to the same object, but Vygotsky believes that the meanings are different. Vygotsky believes that the meanings of words evolve during childhood. Word meanings are dynamic, not static, entities.

Thought is therefore determined by language. And both are determined by society. Language provides a semiotic mediation of knowledge. Language guides the child's cognitive growth. Cognition thus develops in different ways depending on the cultural conditions.

Vygotsky thinks that higher mental functions, too, have social origins.

Language is a system of signs that the individual needs in order to interact with the environment and it is only after this that it is interiorized and can be utilized to express thought. The meaning of a word is initially a purely emotional fact. Only with time will it acquire a precise reference to an object and then an abstract meaning.

Child development is a sequence of stages that lead to the transformation of an interpersonal process into an intrapersonal process.

Children think by memorizing, while adults memorize by thinking. In children something is memorized, in adults the individual memorizes something. In the former case a link is created because of the simultaneous occurrence of two stimuli. In the latter case the individual creates that link. Remembering is transformed into an external activity. Humans are then able to influence their relation with the environment and through that environment change their own behavior. The mastering of nature and the mastering of behavior are interdependent.

Some Nativism

Piaget's model of cognitive development constitutes a powerful paradigm but does not explain everything. In particular, Piaget's theory is inadequate to explain how children learn language. Without any a-priori knowledge of language, it would be terribly difficult to learn the theory of language that every child eventually learns.

The British psychologist Annette Karmiloff-Smith, a student of Piaget, proposed a model of child development that bridges Fodor's nativism (built-in knowledge) and Piaget's constructivism (learning), i.e. innate capacities of the human mind and subsequent representational changes. Karmiloff-Smith envisions a mind that is both equipped with some innate capacities and that grows through a sequence of subsequent changes.

Karmiloff-Smith's child is genetically pre-wired to absorb and organize information in an appropriate format. Each module develops independently, as proved by children who exhibit the symptoms of a single mental disorder but are perfectly capable in all other ways.

Karmiloff-Smith's starting point is Fodor's model of the mind (that the mind is made of a number of independent, specialized modules), but, based on evidence of the brain's plasticity (the brain can restructure itself to adapt to an early damage), Karmiloff-Smith believes that modules are not static and that they "grow" during the child's development, and that new modules are created during the child's development ("gradual modularization").

She points out that children display from the very beginning a whole array of cognitive skills, albeit still unrelated and specific (for example, identifying sounds, imitating other people's movements, recognizing the shapes of faces). Therefore, the child must be born with a set of pre-wired modules that account for these cognitive skills.

Somehow, during development the modules start interacting and working together and adult life takes shape.

Initially, children learn by instinct, or at least "implicitly". Then their thinking develops, and consists of redescribing the world from an implicit form to more and more explicit forms, to more and more verbal knowledge.

Of course, the environment that drives the mind's growth also includes the other individuals. Education and playing are forms of influencing the evolution of the thought system of a child.

Karmiloff-Smith notes a thread that is common to several spheres of cognition: the passage from procedural non-expert to the automatic (nonprocedural) expert also involves a parallel passage from implicit to explicit knowledge (from executing mechanically to understanding how it works). Child development is not only about learning new procedures, it is about building theories of why those procedures do what they do. This "representational redescription" occurs through three stages: first the child learns to become a master of some activity; then she analyzes introspectively what she has learned; and, finally, she reconciles her performance with her introspection. At this point the child has created a "theory" of why things work the way they work.

Therefore Karmiloff-Smith admits cognitive progress like Piaget, but her "representational redescription" occurs when the child has reached a stable state (mastery), whereas in Piaget's model progress only occurs when the child is in a state of disequilibrium.

This process of "representational redescription" involves re-coding information from one representational format (the procedural one) to another (a quasi-linguistic format). There are therefore different levels at which knowledge is encoded (in contrast with Fodor's language of thought).

The same "redescription" process operates within each module, but not necessarily at the same pace. In each field, children acquire domain-specific principles that augment the general-purpose principles (such as representational redescription) that guide their cognitive life.

Moreover, the cultural context determines which modules arise.

Finally, mapping across domains is a fundamental achievement by the child's mind.

Another US developmental psychologist, Patricia Greenfield, modified this theory by showing that initially the child's mind has no modules, only a general-purpose learning system. Modules start developing later in the life of the child. Greenfield identified a common neurological layer in the early stages of development (up to two years old) that accounts for both linguistic skills and object manipulation skills. As the child's brain develops, those skills split.

Adulthood

Expanding on Sigmund Freud's views of child development, in the 1950s the German psychoanalyst Eric Erikson placed emphasis on culture and viewed development as structured in eight stages that extend from birth to death (and not only from birth to late childhood), each stage corresponding to a "task": Infant (in which the task is simply to trust others and the environment), Toddler (in which the task is to to master the physical environment, including learning to walk, talk and eat), Preschooler (in which the task is to imitate adults), School-Age Child (in which the task is to build self-esteem by refining skills), Adolescent (in which the many social roles of son, sibling, student and so forth are integrated mostly through role models and peer pressure), Young Adult (in which personal commitments to friend, lover, spouse, parent and so forth become relevant), Middle-Age Adult (in which the emphasis shifts towards career, family and politics and, in general, being "in charge"), Older Adult (the age of personal loss and intimations of mortality).

All of these models fail to recognize that there is a stage when we begin to observe the developmental stages of children and to realize that we ourselves went through those stages and are still going through stages. This self-reflective stage is what really makes an adult an adult. We are no longer just the object of development: we are a subject that observes younger people's development and wonders about older people's development.

Also, throughout these stages there is an evolution from absorbing culture to spreading culture. Adolescence is the tipping point when spreading culture becomes as important as absorbing culture. Later in life one is much more involved in spreading culture (e.g., raising children or training others) than in absorbing culture.

The Philosophical Baby

Clearly, childhood must serve a purpose, especially human childhood that appears to be so senseless from an evolutionary point of view (no other cub is so helpless in the world). Children spend years practicing a form of mental gymnastics that other species can't do, and that provides the human species with an evolutionary advantage: the ability to map the world and to imagine worlds that don't exist. The US psychologist Alison Gopnik believes that the ability of imagining alternative worlds (of dealing with "counterfactuals") peaks during childhood. Far from being limited to the "here" and "now" as Piaget assumed, children understand how the real world works and then project that knowledge into many other possible worlds. The fact that children don't seem to be capable of distinguishing between reality and fantasy should not be taken as evidence of cognitive limitation but of cognitive power: they are very creative instead of being passive (like adults tend to be).

Because all of this is due to a physically different structure of the brain (mainly in the prefrontal cortex, that doesn't mature until the mid-20s), Gopnik goes as far as to claim that children and adults are two different subspecies of the same species, Homo Sapiens.

Children are also capable of creating "theories of mind", i.e. of understanding the goals and desires of other children and people in general. The way they understand psychology is similar to the way they understand the physical world: they create "psychological counterfactuals", i.e. imaginary companions. With them they rehearse the rules of psychology the same way that they rehearse the rules of physics when they imagine hypothetical worlds. Children are aware that these imaginary companions don't exist, just like they are aware that imaginary worlds don't exist. Children construct maps of both the physical world and of the psychological world. In both cases children first understand the world, then they imagine hypothetical worlds, then they are ready to actually create worlds. In the physical world this translates into action. In the psychological world this translates into dealing with other minds and trying to make them do what we want them to do (which includes the complex interplay of discussions, strategies, and even deceit and lies).

The key to learning is probability theory. Children act just like scientists. They perform experiments and then update their beliefs, each belief being weighed probabilistically. This is true of both facts about the world and facts about people. Gopnik concludes that statistical analysis is actually wired inside the brain. When Thomas Bayes formulated his theorem (the main building block of probabilistic reasoning) he had actually stumbled on a property of the brain.

However, the way children acquire their knowledge about the world is different from the way a scientist does because their brain works in a fundamentally different way. The adult brain regulates attention by inhibiting distractions and thereby "focusing" on something. The baby's brain is still lacking in inhibitory neurotransmitters and therefore absorbs everything that is going on rather than focusing on one particular aspect of reality. The adult brain can balance cholinergic and inhibitory neurotransmitters, whereas the baby's brain is dominated by cholinergic ones.

The Ratchet Effect

German anthropologist Michael Tomasello believes that humans are genetically equipped with the "ability to identify with conspecifics." This attitude is particularly visible in children, who spend most of their time imitating others. Natural selection rewarded the best "copycats". This ability is crucial to the development of social skills such as language: children learn a language because their brains are predisposed to "identify with conspecifics" and because they are exposed to such conspecifics (they socialize with speaking humans). However, this skill is shared by other primates, who nonetheless never attain the sophistication of human civilizations: they only learn by imitation, without having any clue as to why others do what they do. They are are mere copycats.

The key factor (which Tomasello thinks emerges in the ninth month of life) in the development of human civilizations is the understanding of others as goal-directed agents. Tomasello thinks that recognizing the intentions of others is crucial to "learn" from previous generations. Humans do not just imitate other humans: humans also understand why other humans did what they did. Tomasello thinks that this is the secret of rapid learning and of transmission of learned knowledge from one generation to the next one.

Over evolutionary and historical time, the "ratchet effect" due to this attitude to imitation generates the civilizations we are familiar with, civilizations that other primates cannot even dream of. Evolutionarily speaking, the progress made by the human species is impressive. Tomasello believes that the secret to such speedy evolution lies in the unique human attitude towards conspecifics.

The Evolution of Memories

Memory is not a storage device, because it cannot recall events exactly the way they were. Memories change all the time, therefore memory is not a static system, it is a dynamic system.

Memory is pivotal for the entire thought system of the individual. Therefore, memory is about thought, it is not limited to remembering. Memory stores and retrieves thoughts.

Memory can be viewed as an evolving population of thoughts. Thoughts that survive and reproduce are variations of original thoughts, and somehow "contain" those original thoughts, but adapted to the new circumstances. Memories are descendants of thoughts that occurred in the past. Thoughts are continuously generated from previous ones, just like the immune system generates antibodies all the time and just like species are created from previous ones.

Memory, far from being a static storage, is changing continuously. It is not a location, it is the collective process of thinking.

Further Reading
Baddeley, Alan: YOUR MEMORY (MacMillan, 1982)
Baddeley, Alan: WORKING MEMORY (Clarendon Press, 1986)
Baddeley, Alan: HUMAN MEMORY (Simon & Schuster, 1990)
Barsalou, Lawrence: COGNITIVE PSYCHOLOGY (Lawrence Erlbaum, 1992)
Bartlett, Frederic Charles: REMEMBERING (1932) (Cambridge Univ Press, 1967)
Broadbent, Donald: PERCEPTION AND COMMUNICATION (Pergamon, 1958)
Broadbent, Donald: DECISION AND STRESS (Academic Press, 1971)
Bruner Jerome: A STUDY OF THINKING (Wiley, 1956)
Campbell, Joseph: PRIMITIVE MYTHOLOGY: THE MASKS OF GODS (Viking, 1959)
Collins, Alan: THEORIES OF MEMORY (Lawrence Erlbaum, 1993)
Crowder, Robert: PRINCIPLES OF LEARNING AND MEMORY (Erlbaum, 1976)
Eichenbaum, Howard: COGNITIVE NEUROSCIENCE OF MEMORY (Oxford Univ Press, 2002)
Erikson, Eric: IDENTITY AND THE LIFE CYCLE (1959)
Estes, William: CLASSIFICATION AND COGNITION (Oxford University Press, 1994)
Fauconnier, Gilles: MENTAL SPACES (MIT Press, 1994)
Greene, Robert: HUMAN MEMORY (Lawrence Erlbaum, 1992)
Gopnik, Alison: THE PHILOSOPHICAL BABY (Farrar, Straus and Giroux, 2009)
Greenfield, Patricia: LANGUAGE, TOOLS, AND BRAIN (1991)
James, William: PRINCIPLES OF PSYCHOLOGY (1890)
Johnson, Mark: The Body in The Mind (University of Chicago Press, 1987)
Karmiloff-Smith Annette: BEYOND MODULARITY (MIT Press, 1992)
Keil Frank: CONCEPTS, KINDS AND COGNITIVE DEVELOPMENT (Cambridge University Press, 1989)
Lakoff, George: WOMEN, FIRE AND DANGEROUS THINGS (Univ of Chicago Press, 1987)
Lenneberg, Eric: BIOLOGICAL FOUNDATIONS OF LANGUAGE (Wiley, 1967)
Piaget, Jean: EQUILIBRATION OF COGNITIVE STRUCTURES (University of Chicago Press, 1985)
Reiser, Morton: MEMORY IN MIND AND BRAIN (Basic, 1990)

Roediger, Henry: VARIETIES OF MEMORY AND CONSCIOUSNESS (Lawrence Erlbaum, 1989)

Rosch, Eleanor: COGNITION AND CATEGORIZATION (Erlbaum, 1978)

Schacter, Daniel & Tulving Endel: MEMORY SYSTEMS (MIT Press, 1994)

Schacter, Daniel: SEARCHING FOR MEMORY (Basic Books, 1996)

Schanks, David: HUMAN MEMORY: A READER (Oxford Univ Press, 1997)

Semon, Richard: DIE MNEME (1904)

Tolman, Edward: PURPOSIVE BEHAVIOR IN ANIMALS AND MEN (1932)

Tulving, Endel: ORGANIZATION OF MEMORY (Academic Press, 1972)

Tulving, Endel: ELEMENTS OF EPISODIC MEMORY (Oxford Univ Press, 1983)

Tulving, Endel: OXFORD HANDBOOK OF MEMORY (Oxford Univ Press, 2000)

Wittgenstein, Ludwig: PHILOSOPHICAL INVESTIGATIONS (Macmillan, 1953)

MACHINE INTELLIGENCE

The Machinery of the Mind

If i replace every neuron in your brain with an artificial neuron made of silicon that carries out the exact same function, are you still you? If you answer "yes", you just answered "yes" to the question whether machines can be as conscious as humans. This thought experiment is more than just a neuroscience variant on Theseus's paradox that asked whether a ship rebuilt exactly the same with different parts is still the same ship. When it comes to the brain, we're not only asking whether it's still the same brain, but also whether it's still an "i" (a mind, a conscious being) and, more importantly, whether it's still the same person. As chip implants will become common to replace parts of the brain that have been damaged, this will no longer be an academic question: you may have to decide whether to replace a piece of your brain with a piece of silicon that makes your brain work "perfectly well", but is that thing that is "working perfectly well" you? Prosthetic limbs replace limbs that are not responsible for our conscious life. The brain, as far as we know, is. Replacing a piece of the brain with "prosthetic brain regions" implies a belief that our conscious life does not depend on the "stuff" the brain is made of, only on the processes that are going on, in which case one could build an entirely artificial brain that is a conscious being.

The fascination with the idea of building an artificial mind dates from centuries ago. Naturally, before building an artificial mind one should first figure out what kind of machine the human mind is like. The limit to this endeavor seems to be the complexity of the machines we are capable of building. Descartes compared the mind to water fountains, the Austrian psychologist Sigmund Freud to a hydraulic system, the Russian physiologist Ivan Pavlov to the telephone switchboard and the US mathematician Norbert Wiener to the steam engine. Today our favorite model is the electronic computer. Each of these represented the most advanced technology of the time. The computer does represent a quantum leap forward, because it is the first machine that can be programmed to perform different tasks (unlike, say, dishwashers or refrigerators, which can only perform one task).

There is very little similarity between an electronic computer and a brain. They are structurally very different. The network of air conditioning conducts in a high-rise building is far more similar to a brain than the motherboard of a computer. The main reason to consider the electronic computer a better approximation of the brain is functional, not structural: the computer is a machine that can achieve a lot of what a brain can achieve. But that alone cannot be the only reason, as machines capable of representing data and computing data could be built out of biological matter or even crystals. The real reason is still that the computer is the most complex machine we ever built. We implicitly assume that the brain is the most complex thing in the world and that complexity is what defines its uniqueness. Not knowing how it works, we simply look for very complex apparati. Our approach has not changed much since Descartes. We just have a more complex machine to play with.

It is likely that some day a more complex machine will come by, probably built of something else, and our posterity will look at the computer the same incredulous way that today we look at Descartes' water fountains.

Human Logic

The history of Logic starts with the Greeks. Pythagoras's' theorem stands as a paradigm that would influence all of western science: a relationship between physical quantities that is both abstract and eternal. It talks about a triangle, a purely abstract figure which can be applied to many practical cases, and it states a fact that is always true, regardless of the weather, the season, the millennium.

Euclides built the first system of Logic when he wrote his "Elements" (around 350 BC). From just five axioms (there is a straight line between two points, a straight line can be extended to infinite, there is a circle with any given center and radius, all right angles are equal, two parallel lines never meet), he could deduct a wealth of theorems by applying the same inference rules over and over again.

Then, of course, Aristotle wrote his "Organon" and showed that we employ more than one "syllogism" (more than one kind of reasoning). Although he listed several kinds of logical thinking, only three were widely known and eventually became the foundations of Logic. The law of the excluded middle states that an object cannot have both a property and the opposite property (i cannot be both rich and poor). "Modus ponens" states that: if all B's are C's and all A's are B's, then all A's are C's. "Modus tollens" states that: if all B's are C's and no A's are C's then no A's are B's.

After centuries of Roman indifference and of medieval neglect, Logic resumed its course. Studies on logic, from the "Dialectica" of the French philosopher Pierre Abelard (1100 AD) to the "Introductiones Logicam" of the English philosopher William of Shyreswood (1200 AD), had actually been studies on language. Logic was truly reborn with the "Summa Totius Logicae" of another Englishman, William Ockham (1300 AD), who discussed how people reason and learn. Three centuries later Francis Bacon's "Novum Organum" (1620) and Rene' Descartes' "Discours de la

Methode" (1937) hailed the analytic method over the dialectic method and therefore started the age of modern Science. The German mathematician Gottfried Leibniz emphasized the fact that reasoning requires symbols in his "De Arte Combinatoria" (1676) and co-discovered calculus with Isaac Newton. In 1761 the Swiss mathematician Leonhard Euler showed how to do symbolic logic with diagrams. The British philosopher John Stuart Mill tried to apply logic outside of science in his "System of Logic" (1843). Non-numerical algebra was formalized by the British mathematician Augustus De Morgan in "The Foundations of Algebra" (1844).

Another Englishman, George Boole, was so fascinated by the progress of symbolic logic that in "The Laws Of Thought" (1854) he claimed that logic could be applied to thought in general: instead of solving mathematical problems such as equations, one would be able to derive a logical argument. Boole's ideas evolved into "propositional logic" and then "predicate logic", which fascinated philosopher-mathematicians such as Gottlob Frege in Germany, Giuseppe Peano in Italy, Charles Sanders Peirce in the United States, and Bertrand Russell in Britain. Thought became more and more formalized. Frege's "Foundations of Arithmetic" (1884) and "Sense and meaning" (1892), Peano's "Arithmetices Principia Nova Methodo Exposita" (1889), and Russell's "Principia Mathematica" (1903) moved philosophy towards an "axiomatization" of thought.

Formal Systems

David Hilbert, a German mathematician of the beginning of the 20th century, is credited with first introducing the question of whether a mechanical procedure exists for proving mathematical theorems (fully in 1928). His goal was to reduce Mathematics to a more or less blind manipulation of symbols through a more or less blind execution of formal steps. Already implicit in Hilbert's program was the idea that such a procedure could be carried out by a machine. The discipline of formal systems was born, with the broad blueprint that a formal system should be defined by a set of axioms (facts that are known to be true) and a set of inference rules (rules on how to determine the truth or falsity of a new fact, given the axioms). By applying the rules on the axioms, one could derive all the facts that are true.

A formal system employs the language of propositions (statements that can only be true or false and can be combined by binary operators such as "not", "and" and "or") and predicates (statements with a variable that can be quantified existentially or universally, i.e. can be only true or false relative to "at least" one value of the variable or "for every" value of the variable). For example, the fact that Piero Scaruffi is a 51-year old writer could be expressed as: "writer (Piero) AND age (Piero, 51)". The fact that teachers are poor can be expressed with the expression: "FOR EVERY x, teacher(x) -> poor (x)"; that translates as: every individual that satisfies the predicate "teacher" also satisfies the predicate "poor". The fact that some teachers are obnoxious can be expressed as: "FOR AT LEAST ONE x teacher(x) -> obnoxious (x)".

The language of Logic is not very expressive but it lends itself to logical reasoning, i.e. deduction.

This generation of mathematicians basically pushed logical calculus to the forefront of the tools employed to investigate the world. The apparatus of formal systems became the apparatus that one must use to have any scientific or philosophical discussion. Implicit in their program was the belief that the laws of logic "were" the laws of thought.

Incompleteness

Unfortunately, a number of logical paradoxes refused to disappear, no matter how sophisticated Logic became. All of them, ultimately, are about self-reference.

Oldest was the liar's paradox: the sentence "I am lying" is true if false and false if true. Bertrand Russell came up with the brilliant paradox of the class of classes that do not belong to themselves: such a class belongs to itself if it does not belong to itself and viceversa. This paradox is also known as the paradox of the barber who shaves all barbers who do not shave themselves (does he shave himself?). A variation on these paradoxes is often used to prove the impossibility of an omnipotent God: if God can do anything, can he build a rock that is so heavy that even s/he cannot lift it?

Hilbert was already aware of these paradoxes, but several proposals had been made to overcome them. And several more will be proposed later (from Bertrand Russell's "Theory of Types" of 1908 to Jon Barwise's "Situation Theory" of 1986).

Nevertheless, Hilbert and others felt that Logic was capable of proving everything. Hilbert's goal was to find the procedure that would solve all possible problems. By applying that procedure, even non-mathematicians would prove difficult mathematical theorems. Hilbert was aware that, by applying inference rules of Logic to the facts that are known to be true, one could list all the other facts that follow to be true. The power of Logic seemed to be infinite. It made sense to imagine that Logic was capable of proving anything.

The dream of a purely mechanical procedure for solving mathematical problems was shattered by yet another paradox, the one known as Goedel's theorem. In 1931 the (Czech-born) Austrian mathematician Kurt Goedel proved that any formal system (containing the theory of numbers, i.e.

Arithmetic) contains a proposition that cannot be proven true or false within that system (i.e., an "undecidable" proposition).

Intuitively, Goedel's reasoning was that the statement "I cannot be proven" is true if and only if it cannot be proven; therefore in every system there is always at least one statement that cannot be proven, the one that says "I cannot be proven".

Neither the proposition nor its negation can be proven within the system. We can't know whether it is true or false. Predicate Logic, for example, is undecidable. Therefore any formal system built on Predicate Logic happens to be built on shaky foundations. And that includes pretty much all of classical Logic. The conclusion to be drawn from Goedel's theorem is catastrophic: the very concept of truth cannot be defined within a logical system. It is not possible to list all the propositions that are true.

Hilbert had reduced Logic to a mechanical procedure to generate all the propositions that can be proven to be true in a theory. The dual program, of reducing Logic to a mechanical procedure to prove theorems (to prove if a proposition is true), is impossible because of Goedel's theorem (since it is not always possible to prove that a proposition is true). This came to be known as the "decision problem".

Note that Hilbert was looking for an "algorithm" (a procedure), not a formula. For centuries most of science and Mathematics had focused on formulas. The mighty apparatus of Physics was built on formulas. All natural sciences were dealing with formulas. Hilbert, indirectly, started the trend away from formulas and towards algorithms, a trend that would become one of the salient leitmotivs of this century. A formula permits to compute a result directly from some factors by applying mathematical operations in the sequence prescribed by the priority rules of operators. An algorithm prescribes a step-by-step procedure for achieving the result. A formula is one line with an equal sign. An algorithm is made of finite steps, and the steps are ordered. Each step can be a mathematical operation or a comparison or a change in the sequence of steps. Each step can be conceived of as an "instruction" to an ideal machine capable of carrying out those elementary steps.

Truth and Meaning

One more notion was necessary to complete the picture: meaning. What did all this mean in the end?

Aristotle had realized the importance of "truth" for logical reasoning and had offered his definition: a proposition is true, if and only if it corresponds with the facts. This is the "correspondence theory of truth".

Frege had founded Logic on truth: the laws of logic are the laws of truth. Truth is Frege's unit of meaning. In fact, it was Frege who introduced "true" and "false", the so called "truth values". Frege regarded logical propositions as expressing the application of a concept to an object, as in "author(piero)" that states that Piero is an author. Indirectly, he partitioned the universe into concepts and objects, equated concepts with mathematical functions and objects with mathematical terms. The proposition "Piero is an author" has a concept "author" that is applied to a term "Piero". All of this made sense because, ultimately, a proposition was either true or false, and that could be used to think logically.

According to Aristotle, if this proposition is true, then its meaning is that the person referred to as Piero is an author; and viceversa.

Hilbert had taken this course of action to its extreme consequences. Hilbert had emancipated Logic from reality, by dealing purely with abstractions.

In 1935 the Polish mathematician Alfred Tarski grounded Logic back into reality. He gave "meaning" to the correspondence theory of truth.

Logic is ultimately about truth: how to prove if something is true or false. But what is "truth"? Tarski was looking for a definition of "truth" that would satisfy two requirements, one practical and one formal: he wanted truth to be grounded in the facts, and he wanted truth to be reliable for reasoning. The second requirement was easily expressed: true statements must not lead to contradictions. The first requirement was more complicated. How does one express the fact that "Snow is white" is true if and only if snow is white? Tarski realized that "snow is white" is two different things in that sentence. They are used at different levels. A proposition p such as "snow is white" means what it states. But it can also be mentioned in another sentence, which is exactly the case when we say that "p is true". The fact that "p" is true and the sentence "p is true" are actually two different things. The latter is a "meta-sentence", expressed in a meta-language. In the meta-language one can talk about elements of the language. The liar's paradox, for example, is solved because "I am lying" is a sentence at one level and the fact that I am telling the truth when I am lying is a sentence at a different level; the contradiction is avoided by considering them at two different levels (language and meta-language).

Tarski realized that truth within a theory can be defined only relative to another theory, the meta-theory. In the meta-theory one can define (one can list) all the statements that are true in the theory.

Tarski introduced the concepts of "interpretation" and "model" of a theory. A theory is a set of formulas. An interpretation of a theory is a function that assigns a meaning (a reference in the real world) to each of its formulas. Every interpretation that satisfies all formulas of the theory is a model for that theory. For example, the formulas of Physics are interpreted as laws of nature. The universe of physical objects becomes a model for Physics. Ultimately, Tarski's trick was to build "models" of the world which yield "interpretations" of sentences in that world. The important fact is that all semantic concepts (i.e., meaning) are defined in terms of truth, and truth is defined in terms of satisfaction, and satisfaction is defined in terms of physical concepts (i.e., reality). The meaning of a proposition turns out to be the set of situations in which it is true.

What Tarski realized is that truth can only be relative to something. A concept of truth for a theory (i.e., all the propositions that are true in that theory) can be defined only in another theory, its "meta-theory", a theory of that theory. All paradoxes, including Goedel's, can then be overcome, if not solved. Life goes on.

Tarski grounded meaning in truth and in reference, a stance that would set the stage for a debate for the rest of the century.

The Turing Machine

The two great visionaries of computation, the British mathematician Alan Turing and the Hungarian mathematician John Von Neumann, had a number of influential ideas. Among the many, in 1936 Turing formalized how a machine can perform logical calculus ("On computable numbers, with an application to the Entscheidungsproblem", 1936), and, a few years later, Von Neumann explored the possibility that a machine could be programmed to make a copy of itself. In other words, Turing laid the foundations for the discipline of building an "intelligent" machine, and Von Neumann laid the foundations for the discipline of building a self-reproducing machine.

Turing defined computation as the formal manipulation of symbols through the application of formal rules (Hilbert's view of Logic), and devised a machine ("On Computable Numbers, with an Application to the Entscheidungsproblem", 1936) that would be capable of performing any type of computation (Hilbert's dream).

Hilbert's ideas can be expressed in terms of mathematical "functions". A predicate can always be made to correspond to a function. For example, "age(Person,Number)" corresponds to the function "Number= age(person)"; and viceversa. Both mean that the age of Person is Number. The advantage of using functions instead of predicates is that it is easier to manipulate functions than predicates. For example, the US mathematician Alonzo Church showed how two functions can be compared. A function can be defined in an "extensional" way (the pairs of input and output values) or in an "intensional" way (the computational procedure it performs). Comparing two extensional definitions can take forever (there can be infinite pairs of input and output). In order to compare two intensional definitions, Church invented the "Lambda abstraction", which provides rules to transform any function in a "canonical" form. Once they are in canonical form, two functions can be easily compared.

In the language of functions, Hilbert's goal was to find a mechanical procedure to build all computable (or "recursive") functions. So a recursive function is defined by an algorithm, which in turn can be implemented by a computer program. Recursive functions correspond to programs of a computer. Not surprisingly, it turns out that a predicate is decidable (can be proven true or false) if and only if the corresponding function is recursive, i.e. computable.

Turing realized this and, when he set himself to find a mechanical procedure to perform logical proofs, he basically set himself to invent the computer (at least conceptually). His thought experiment, the "Turing Machine", is the algorithm that Hilbert was looking for, and it turns out that it is also the general algorithm that fuels electronic computers.

A Turing Machine is capable of performing all the operations that are needed to perform logical calculus: read current symbols, process them, write new symbols, examine new symbols. Depending on the symbol that it is reading and on the state in which it is, the Turing machine decides whether it should move on, turn backwards, write a symbol, change state or stop. Turing's machine is an automatic formal system: a system to automatically compute an alphabet of symbols according to a finite set of rules.

Church argued that everything that is computable in nature can be computed with a Turing machine.

But there can be infinite Turing machines, depending on the rules to generate new symbols. So Turing described how to build a machine that would simulate all possible Turing machines.

The "Universal Turing Machine" is a Turing Machine capable of simulating all possible Turing Machines. It contains a sequence of symbols that describes the specific Turing machine that must be simulated. For each computational procedure, the universal machine is capable of simulating a machine that performs that procedure. The universal machine is therefore capable of computing any computational function. In other words, since the Universal Turing Machine is a machine capable of performing any other machine, it is capable of solving all mathematical problems. A computer is nothing but a Turing machine with a finite memory.

When Von Neumann neatly divided the data from the instructions (instructions are executed one at the time by the "processor" and they operate on data kept in a "memory"), he simply interpreted for engineers the concepts of the Universal Turing Machine: a computer (the hardware) can solve any problem if it is fed the appropriate program (the software).

That architecture was to become the architecture of the computer and today's "sequential" computers (the most common varieties) are still referred to as "Von Neumann architectures". They are "sequential" in the sense that they are controlled by software programs and execute one instruction after the other from those programs.

Ultimately, Turing reduced Hilbert's program to manipulation of symbols: logic is nothing more than symbol processing. Indirectly, he turned the computer into the culminating artifact of Hilbert's formal program.

Turing showed that rational machines are feasible. Furthermore, he showed that one can build a rational machine that can perform "any" rational task.

As for Hilbert's decision problem, both Church and Turing had basically offered a definition of "algorithm" (one based on Lambda Calculus and the other one based on the Turing machine), both definitions being equivalent and both showing that Hilbert's question could not be answered: there is no universal algorithm to solve every mathematical problem, or there is no universal algorithm for deciding whether or not a Turing machine will stop. Turing's argument was very similar to Goedel's incompleteness theorem whereas Alonzo Church had already come up with his own proof that first order logic is undecidable ("An unsolvable problem of elementary number theory", 1935).

Later the Israeli physicist David Deutsch generalized Turing's ideas and envisioned a "quantum" machine in which Turing states can be linear combinations of states ("Quantum Theory, The Church-Turing Principle And The Universal Quantum Computer", 1985). The behavior of a quantum machine is a linear combination of the behavior of several Turing machines. A quantum machine can only compute recursive functions, just like Turing's machine, but it turns out to be much faster in solving problems that exhibit some level of parallelism. In a sense, a quantum computer is capable of decomposing a problem and delegating the sub-problems to copies of itself in other universes.

Cybernetics

Cybernetics is the science of control and communication.

Cybernetics was born out of the passion and excitement generated by the spread of complex mechanical and electrical machines, whose functioning was largely based on control processes. Cybernetics found similarities between some of those mechanical processes and some biological processes.

The concepts introduced by cybernetics built a bridge between machines and nature, between "artificial" systems and natural systems. For example, most machines employ one type of "feedback" or another. Feedback, by sending back the output as input, helps control the proper functioning of the machine.

The self-regulatory character of the human nervous system had been emphasized since the 1920s by the Russian physiologist Nikolai Bernstein. He concluded that, given the complexity of the human motor system, movement could not possibly be "commanded" by a central processor. Instead, he thought that movement was achieved by continually analyzing sensory inputs and adjusting motor output consequently. When I extend the arm to grab something, I have not computed exactly the trajectory and speed of my movement. I am re-computing it every second as my arm approaches the object. There is no computer in the brain that calculates the exact trajectory for the arm and the hand and the fingers in order to reach the glass. There is a continuous dialogue between the senses and the arm, the hand and the fingers, so that the trajectory is adjusted as motion proceeds. Both machines and living beings rely on "control systems".

Feedback

"Negative" feedback occurs when the output of the engine is fed back into the engine for the purpose of "controlling" it. For example, every engine has a valve that helps stabilize its power: the valve opens or closes depending on whether the engine is working too little or too much. The resulting power is always the same because the valve "balances" the work of the engine. The valve does so by opening or closing in a manner that depends on the work of the engine: in other words, the output of the engine is used to determine how much of the output of the engine has to be curtailed. This is negative feedback because the valve operates "against" the engine: it reverses the trend of the engine. The valve is canceling the fluctuations in the work of the engine.

"Positive" feedback occurs when those fluctuations are amplified, not canceled. The output of the engine is fed back into the engine for the purpose of reinforcing it. Instead of a stable output, we get runaway acceleration or complete rest, because positive feedback increases a perturbation instead of curbing it. Needless to say, positive feedback is not often used by engineers, who are more interested in building stable machines, rather than machines that rapidly self-destroy. But positive feedback is common in nature, where it determines the size of a population (until negative feedback prevails in the form of limited resources) and aggressive behavior (until negative feedback prevails in the form of a stronger opponent).

In human societies positive feedback is not rare. For example, positive feedback is often responsible for bestsellers: a record will sell more once it enters the best-selling charts and it will keep getting more popular for the simple reason that it is popular.

Homeostasis

The US mathematician Norbert Wiener (who co-founded Cybernetics in 1943) recognized the importance of feedback (a term that he coined) for any meaningful behavior in the environment: a system that has to act in the environment must be able to continuously compare its performed action with the intended action and then infer the next action from their difference. This is what all living organisms do all the time in order to survive.

Feedback is the action of feeding a system its past performance. Given past performance, the system can adjust future performance. All biological systems (animals, plants, ecosystems) exhibit feedback. Feedback is the basis of life. As Bernstein had asserted, we could not even coordinate our limbs if we were not capable of using feedback.

Feedback is crucial for "homeostasis", the phenomenon, first described by the US biologist Walter Cannon ("Organization for physiological homeostasis", 1929), by which an organism tends to compensate variations in the environment in order to maintain its internal stability; i.e., by which an organism adapts to the environment. Homeostasis consists in maintaining a constant internal state in reaction to changes in the environment and through action on the environment. For example, body temperature is controlled by perspiring and shivering (sweat lowers the internal temperature, shivers increase it). Homeostasis is crucial for survival.

Given the number of factors that must be taken into account for any type of action in the real world, it is not surprising that the brain evolved to use feedback, rather than accurate computation, to guide motion.

It is not a coincidence that feedback turns out to be as crucial also for the performance of machines in their environment. Beginning with James Watt's steam engine, machines have been designed so as to be able to control themselves.

A control system is a system that uses feedback to achieve some kind of steady state. A thermostat is a typical control system: it senses the temperature of the environment and directs the heater to switch on or off; this causes a change in the temperature, which in turn is sensed by the thermostat; and so forth. This loop of action and feedback, of sensing and controlling, realizes a control system. A control system is therefore capable of achieving a "goal", is capable of "purposeful" behavior (as opposed to the chaotic behavior that would result if feedback was not used).

Living organisms are control systems. Most machines are also control systems.

Another folk concept that Wiener formalized is "noise". Wiener emphasized that communication in nature is never perfect: every message carries some involuntary "noise" and in order to understand the communication the original message must be restored. This led to a statistical theory of amount of information.

Wiener understood the essential unity of communication, control and statistical mechanics, which is the same whether the system is an artificial system or a biological system. This unified discipline became "Cybernetics". A cybernetic system is a system that achieves an internal homeostatic control through an exchange of information between its parts.

Alfred Russel Wallace predated cybernetics by a century when ("On the Tendency of Varieties to depart indefinitely from the Original Type", 1858) he noticed that the principle of natural selection is a self-controlling system similar to the principle of the steam engine.

The British neurologist Ross Ashby also placed emphasis on feedback ("Principles of the self-organizing dynamic system", 1947). Both machines and living beings tend to change in order to compensate variations in the environment, so that the combined system is stable. For living beings this translates into "adaptation" to the environment. The "functioning" of both living beings and machines depends on feedback processes to the extent that feedback allows the system to self-organize. Self-organizing systems are systems made of a very high number of simple units which can evolve autonomously and adapt to the environment by virtue of their structure. Basically, each subsystem adapts to the environment created by the other subsystems, and the result of this collective adaptation is self-organization into a stable state.

Ashby believed that in every isolated system that is subject to constant forces "organisms" arise that are capable of adapting to their environment (i.e., that tend towards stationary or quasi-stationary non-equilibrium states). Indirectly, his principle of self-organization asserts that in any isolated system, life and intelligence inevitably develop.

Ashby also argued that a system (whether biological or mechanical) cannot create anything new unless it contains a source of randomness.

Control Systems

The US electrical engineer William Powers extended these ideas to a hierarchical organization of control systems. First of all, he realized that a control system controls what it senses: it controls its input (the perception), not its output (the behavior). A thermostat controls the temperature, not the gas consumed by the heater. Organisms change their behavior, but they do it in order to control a perception. Behavior is the control of perception.

A control system is a blind performer of a specific function. It does not know the larger scope of its function. For example, a thermostat does not know why it is keeping the temperature constant. A control system has an internal goal (e.g., to maintain temperature constant) and its behavior is determined by the difference between what it perceives and its internal goal. Nonetheless, the control system exhibits a behavior that appears to be "purposeful".

Next, he envisioned a system which is made of a pyramid of control systems, each one sending its output to some "lower-level" control systems. The lowest level in the hierarchy is made of control systems that use sensors to sense the environment and "effectors" to act on the environment, and some "reference level" to determine what they have to maintain at a constant level. For example, a thermostat would sense the environment's temperature, effect the heater and maintain the measured temperature at a constant level. At a higher level, a control system senses and effects the reference level of lower-level control systems. An engine could direct a thermostat to maintain a certain temperature. The reference level of the lower level is determined by the control systems of the higher level.

Living organisms are made of such hierarchies of control systems. "Instinctive" behavior is the control system (organized in a hierarchy) that the organism inherits at birth. They determine internally what parameters have to be maintain constant, and at which magnitude. Behavior is a backward chain of behaviors: walking up the hierarchy one finds out why the system is doing what it is doing (e.g., it is keeping the temperature at such a level because the engine is running at such a speed because… and so forth). The hierarchy is a hierarchy of goals (goals that have to be achieved in order to achieve other goals in order to achieve other goals in order to…)

This hierarchy inevitably extends outside the system and into the environment. A machine is part of a bigger machine which is part of a factory which is part of an economy which is part of a society which is part… The goal of a component of the machine is explained by a chain of higher-level goals that extend into society. In the case of living organisms, the chain of goals extends to their ecosystem, and ultimately to the entire system of life.

Algorithms and Automata

Cybernetics implied a paradigm shift from the world of continuous laws to the world of algorithms. Physical sciences had been founded on equations that were continuous, but Cybernetics could not describe any feedback-based process with one continuous equation. The most natural way to describe such a process was to break it down into the sequence of its constituent steps, one of which refers ("feeds back") to a previous one. Every mechanical process could then be interpreted as a sequence of instructions that the machine must carry out. Indirectly, the complex clockwork of a watch is carrying out the sequence of instructions to compute the time. The watch is, in a sense, an automaton that performs an algorithm to compute the time.

The effect of an algorithm is to turn time's continuum into a sequence of discrete quanta, and, correspondingly, to turn an analog instrument into a digital instrument. A watch, for example, is the digital equivalent of a sundial: the sundial marks the time in a continuous way, the watch advances by seconds.

The digital world (of discrete quantities) differs from the analog world (of continuous quantities) in a fundamental way when it comes to precision. An analog instrument can be precise, and there is no limit to its precision. A digital instrument can only be approximate, its limit being the smallest magnitude it can measure (seconds for a watch, millimeters for a ruler, centigrades for a thermometer, etc.). For the purpose of "recognizing" a measurement, though, a digital reading is often better: while two analog values can be so close that they can be confused, two digital values are unambiguously either identical or different. In the context of continuous values, it is difficult to decide whether a value of 1.434 and a value of 1.435 should be considered as the same value with a little noise or two different values; whereas in the context of binary values (the binary universe being a special case of digital universe), a value is unambiguously either zero or one. This feature has been known even before compact discs replaced vinyl records (the Morse code was an early application of the concept). An analog instrument will probably never measure a one as a one or a zero as a zero (it will yield measurements that are very close to one or very close to zero), whereas a digital instrument cannot measure anything else than a zero or a one because its scale does not have any other value (e.g., a digital watch cannot measure 0.9 seconds because its scale is in seconds). This limitation often translates into an advantage.

What is implicit in a cybernetic scenario is that the world is driven by algorithms, rather than by continuous physical laws. Similar conclusions were reached in Linguistics (a "generative grammar" is run by an algorithm) and in Cognitive Science (a production system is run by an algorithm).

An algorithm is a deterministic process in the form of a sequence of logical steps. A computer program simply implements an algorithm. Reducing the laws of nature to algorithms is like reducing nature to a world of automata.

Information Theory

The Hungarian physicist Leo Szilard, trying to solve the paradox of "Maxwell's demon" (a thought experiment in which measurements cause a decrease of entropy), calculated the amount of entropy generated as the demon stores information in its memory ("On the Decrease in Entropy in a Thermodynamic System by the Intervention of Intelligent Beings", 1929), thereby establishing a connection between information and entropy: information was shown to increase when entropy decreases and viceversa. The tendency of systems to drift from the low-probability state of organization and individuality to the high-probability state of chaos and sameness could be interpreted as a decline in information.

Wiener conceived of information as the opposite of entropy. To him the amount of information in a system was a measure of its degree of organization. Hence, the entropy of a system was a measure of its degree of disorganization. The higher the entropy the lower the information (technically, information is a negative logarithm whereas entropy is a positive logarithm). A process that loses information is a process that gains entropy. Information is a reduction in uncertainty, i.e. of entropy: the quantity of information produced by a process equals the amount of entropy that has been reduced.

An unlikely unusual message is a state of low entropy because there are relatively few ways to compose that message. Its information, however, is very high, precisely because it is so unusual.

The second law of Thermodynamics, one of the fundamental laws of the universe, responsible for our dying among other things, states that an isolated system always tends to maximize its entropy (i.e., things decay). Since entropy is a measure of the random distribution of atoms, maximizing it entails that the distribution has to become as homogeneous as possible. The more homogeneous, the less informative a distribution of probabilities is. Therefore, entropy, a measure of disorder, is also a measure of the lack of information.

The French physicist Leon Brillouin coined the term "negentropy" for Wiener's negative entropy and formulated the "negentropy principle of information" ("Negentropy Principle of Information", 1953): since the total change in the entropy has to be greater than or equal to zero, new information in a system can only be obtained at the expense of the negentropy of some other system.

The US electrical engineer Claude Shannon and the US mathematician Warren Weaver, instead, defined entropy as the statistical state of knowledge about a question: the entropy of a question is related to the probability assigned to all the possible answers to that question.

Shannon's entropy measures the uncertainty in a statistical ensemble of messages. That entropy "is" information. The amount of information is equal to entropy. This is exactly the opposite of Wiener's definition of information.

Shannon defines information as chaos (entropy), Wiener and Brillouin define information as order (negentropy).

In summary, a theory of information turns out to be related to a theory of entropy: if information is ultimately a measure of order, entropy is ultimately a measure of disorder, and, indirectly, a measure of the lack of information; if information is ultimately a measure of chaos, then it is basically entropy. Either way, there is a direct connection between the two.

In this view the role of (positive or negative) entropy was to contribute to the self-organization and increased complexity of a system, an interpretation that constituted a conceptual revolution.

The relationship between information and life had already been understood by the Russian mathematician Aleksandr Lyapunov ("The General Problem of the Stability of Motion", 1892) who had realized that life is about information, and the preservation of life is about processing information. His definition of life reads: "a highly stable state of matter, utilizing information encoded by the states of the individual molecules for the purpose of developing reactions aimed at self-preservation".

Whether unifying machine processes and natural processes, or unifying quantities of Information Theory and quantities of Thermodynamics, the underlying theme was that of finding commonalities between artificial systems and natural systems. This theme caught up speed with the invention of the computer.

Algorithmic Information Theory

A fundamental step in bridging the analog and the digital world was taken in 1933 when the Russian engineer Vladimir Kotelnikov discovered the "Nyquist–Shannon" sampling theorem: how to convert continuous signals (e.g. the wave of a sound) into discrete sequences of numbers (e.g. a sequence of zeroes and ones) in such a way that the original signal can be reconstructed without losing fidelity, That was the fundamental mathematical artifice that made the digital revolution possible. No information is lost if a signal is sampled at the proper frequency.

Another Russian mathematician, Andrei Kolmogorov, formulated "Algorithmic Information Theory" ("On Tables of Random Numbers", 1963), a scientific study of the concept of complexity. Complexity is basically defined as a quantity of information, which means that Algorithmic Information Theory is the discipline that deals with the quantity of information in systems.

The complexity of a system is defined as the shortest possible description of it; or, equivalently, the least number of bits of information necessary to describe the system. It turns out that this means: "the shortest algorithm that can simulate it"; or, equivalently, as the size of the shortest program that computes it. For example, the complexity of "pi" is the ratio between a circumference and its diameter. The emphasis is therefore placed on sequences of symbols that cannot be summarized in any shorter way. Algorithmic Information Theory looks for the shortest possible message that encodes everything there is to know about a system. Objects that contain regularities have a description that is shorter than themselves.

Algorithmic Information Theory represents an alternative to Probability Theory when it comes to study randomness. Probability Theory cannot define randomness. Probability Theory says nothing about the meaning of a probability: a probability is simply a measure of frequency. On the other hand, randomness can be easily defined by Kolmogorov: a random system is one that cannot be compressed. A random sequence is one that cannot be compressed any further.

The Argentinean mathematician Gregory Chaitin proved that randomness is pervasive. His "Diophantine" equation contains 17,000 variables and a parameter which can take the value of any integer number. By studying it, Chaitin achieved a result as shocking as Goedel's theorem: there is no way to tell whether, for a specific value of the parameter, the equation has a finite or infinite number of solutions. That means that the solutions to some mathematical problems are totally random. Chaitin defined randomness in terms of computability: if it is computable (if there is a program that generates it), a number is not random. Conversely, one can measure the degree of randomness of something by the length of the shortest program (algorithm) that generates it.

Incidentally, every system has a finite complexity because of the Bekenstein bound. In Quantum Theory the Bekenstein bound (named after the Israeli physicist Jacob Bekenstein) is a direct consequence of Heisenberg's uncertainty principle: there are upper limits on the number of distinct quantum states and on the rate at which changes of state can occur. In other words, the principle of uncertainty indirectly sets an upper limit on the information density of a system, and that upper limit is expressed by the Bekenstein bound.

Physicists seem to be fascinated with the idea of quantifying the complexity of the brain and even the complexity of a human being. The US mathematician Frank Tipler estimated the storage capacity of the human brain at 10 to the 15th power and the maximum amount of information stored in a human being at 10 to the 45th power (a number with 45 zeros). Freeman Dyson computed the entropy of a human being at 10 to the 23rd.

Artificial Intelligence

The term "Artificial Intelligence" was coined around 1955 by the US mathematician John McCarthy, but it has never been clarified what it was truly supposed to mean. The reason is simple: there is no consensus on what makes a machine (or, for that matter, a human being) "intelligent".

If opinions vary on whether Artificial Intelligence is feasible or not, opinions are even more varied on how Artificial Intelligence should be achieved.

At the beginning Artificial Intelligence was often equated with the quest for the "general problem solver", the program capable of solving all mathematical problems. Because the computer is a symbolic processor, and proving theorems is about processing symbols, it was natural to assume that a computer can prove all theorems. However, scientists soon realized that problem solving is not everything, and in everyday life we can solve problems that are essential to our survival (such as deciding when to cross a street) without ever using the Mathematics we studied in school.

Thus "intelligence" is not commonly defined by the number of theorems one can prove in a second (otherwise machines would already be far more intelligent than the most intelligent humans) but by the ability to move around in the real world and carry on all the tasks that humans carry out more or less effortlessly during the day.

A more realistic view is that intelligence is the result of reasoning about knowledge. Intelligent behavior originates from a base of knowledge and from the ability to carry out inferences on that knowledge base. Intelligence is essentially knowledge processing. Since a computer is ultimately a symbol processor, the issue is then how to express knowledge in a symbolic form.

The difference between knowledge and information is crucial. Information can be found in books, knowledge comes from experience. Common sense, for example, is a form of knowledge but not a form of information. Anybody can access the information stored in a medical encyclopedia, but only physicians have real knowledge about medicine. The focus of Artificial Intelligence is not in building encyclopedias, in storing huge amounts of information: it is in "cloning" humans who are experts (i.e., have acquired specialized knowledge) in a field or domain. The difference between information and knowledge is, for example, the difference between asking "who is the president of the United States?" and asking "who will be the next president of the United States?" The former question requires only "information" about who is the current president, the latter question requires "knowledge" about the domain of politics.

According to John McCarthy ("Programs with Common Sense", 1958), knowledge representation must satisfy three fundamental requirements: "ontological" (must allow one to state the relevant facts), "epistemological" (allow one to express the relevant knowledge) and "heuristic" (allow one to perform the relevant inference). Artificial Intelligence can then be defined as the discipline that studies what can be represented in a formal manner (epistemology) and computed in an efficient manner (heuristics). The language of Logic satisfies those requirements: it allows us to express everything we know and it allows us to make computations on what is expressed by it. Each set of knowledge is in fact a mathematical theory.

The underlying assumption of the knowledge-based approach is that symbolic processing per se may lead to human-like intelligence.

Knowledge Representation

One of the crucial steps to build intelligent machines is therefore knowledge representation: first and foremost, one must encode in a machine the knowledge about the world possessed by humans. Every science needs to build a mathematical model of its world before it can perform any inference and draw any conclusions. Physics, for example, represents natural laws with formulas. Then formulas can be combined to yield prescriptions about the effects of actions. The world of Artificial Intelligence is the world of knowledge: what must be represented formally is knowledge.

Knowledge has traditionally been formalized in three forms: facts, stimulus-response pairs (or cause-effect, or premise-action, or antecedent-consequent pairs), and relations between concepts. Facts are easily represented in first-order Predicate Logic in the form of logical expressions: "Piero is a writer" can be represented as "writer (Piero)", meaning that Piero satisfies the predicate "writer" (or that Piero belongs to the set of individuals that satisfy the predicate "writer"). For example, if we know that all writers are creative, then we can apply a simple step of deduction and derive that Piero is also creative.

"Production" rules are usually employed to express the causal connection between one fact and another fact (if something is true, then something else must be true too). For example, if somebody is a human being, then she is also a mammal. Whenever the antecedent is true, the consequent is also true. This too can be translated into Predicate Logic, because the "implication" is mathematically equivalent to a logical expression (in Logic, p IMPLIES q is equivalent to NOT p OR q). More rules can therefore be combined according to Predicate Calculus.

Finally, relations between concepts (i.e., complex concepts) can be represented with systems such as "semantic networks" and "frames". A semantic network represents concepts as nodes, "links" a concept with other concepts, and specifies of what type each link is. For example, the concept of a human being is linked to the concept of a mammal by a link of type "BELONGS TO". A concept may have many links of many types to other concepts. Ideally, all human knowledge could be represented by a gigantic semantic network.

A "frame" can be used to represent the inner structure of a concept: its attributes, their default values, the actions associated with the attributes, and, again, the links to other concepts. A car's attributes include that its function is to move, that it has four wheels, that it costs so much, etc. Both semantic networks and frames can also be reduced to expressions of first-order Predicate Logic.

Anything that can be reduced to Predicate Logic satisfies McCarthy's requirements.

Expert Systems

An "expert system" is simply a software system that has a knowledge base and some inference methods that can be applied to that knowledge base.

The knowledge base describes the rules that apply to the domain of expertise. The "inference engine" is capable of inferring from those rules the appropriate action in the face of a specific situation. The combination of a knowledge base and an inference engine should therefore yield a machine that behaves just like a human expert (i.e., one who makes the same decisions in the same circumstances) within the domain of expertise represented in the knowledge base.

Since all methods commonly employed to represent knowledge can be reduced to some variant of Predicate Logic, Logic can provide the inference techniques required to draw conclusions from the knowledge base. For example, some representation systems (the "production systems") simply encode knowledge in production rules, and production rules basically assert a new fact within a knowledge base whenever some other facts have been asserted. In presence of a new situation (i.e., of a set of new facts), a number of production rules will "fire" and assert another set of new facts, which in turn will trigger more production rules, and so forth recursively ("forward chaining"). Viceversa, one can prove the truth of a statement by looking up which production rules would assert it and what has to be true in order for them to fire, and so forth recursively ("backward chaining"). This way of reasoning belongs to deduction, the most studied and reliable form of logical reasoning.

An influential paradigm was introduced by the US economist Herbert Simon and the US mathematician Allen Newell. Just as the rules of grammar provide a simple means to generate all possible sentences of a given language, the recursive application of rules can provide a simple means to generate all possible actions in a given "space". The issue then shifts to finding out which particular sequence of actions leads to a solution: the machine has to be able to "search" for that correct sequence. Thus the process performed by an expert system can also be viewed as a "search" in a space of all possible solutions.

Each logical step corresponds to a step in the search through that abstract space for the solution to the current problem. The search can be "blind" or "heuristic": the former recursively applies a set of algorithms (the same ones regardless of the type of problem at hand, such as "modus ponens" or "reductio ad absurdum"), hoping that eventually it will stumble into the solution; the latter employs "clues" about the problem at hand (or "domain heuristics") in order to find short-cuts. The algorithms employed during a heuristic search can be either "weak" methods, such as "hill climbing" and "means-end analysis", which are relatively independent of the domain, and methods which are entirely domain-specific.

The template of an expert system was outlined by Newell's and Simon's "General Problem Solver" (1957), a system designed to solve problems in two steps: first the blind generation of possible solutions and then selection of the solutions that work. It employed a "blackboard" architecture, a model of reasoning in which different agents can post their partial solutions on a common area that other agents can access.

Pioneering expert systems include: Edward Feigenbaum's "Dendral" (1965) for analyzing chemical compounds, Bruce Buchanan's "Mycin" (1972) for diagnosing diseases, John McDermott's "Xcon" (1980) for configuring computers. Since the 1980s a growing number of them entered the workforce. But they were far from exhibiting any "intelligence", other than what one expects from machines.

Note that Newell and Simon had basically mechanized psychology, they had mechanized the very self that (in humans) thinks, searches and finds solutions. Indirectly, their architecture implied that the conscious self is an epiphenomenon, a side effect, an outcome and not a cause, of intellligent behavior.

Programs That Learn: Induction
Intelligence, though, is often regarded as "learning" knowledge rather than simply using it. One of the most cogent failures of the field of expert systems lies in the inability of such systems to learn on their own the knowledge they need in order to operate. The performance of a human being increases with experience, and the rate of increase is often considered a measure of the person's intelligence. In an expert system, performance changes (and does not necessarily improve) only when a new knowledge base is installed. While the intelligence of an expert system is purely deductive, human intelligence is also inductive: new knowledge is continuously inferred.

In order to be capable of "learning", an expert system should continuously change its knowledge base to reflect the outcome of its actions.

Learning can be done according to two opposite paradigms: an inductive paradigm and an analytic (or deductive) paradigm.

The former constructs the symbolic description of a concept from a set of positive and negative instances (instances that belong and instances that do not belong to that concept). Usually, the symbolic description is in the form of a "discrimination" rule: if a new instance satisfies such rule, then it does belong to the concept. This view goes back to the US psychologist Jerome Bruner's theory of concepts: a concept is defined by a set of features which are individually necessary and jointly sufficient for an instance to belong to that concept. The corresponding algorithm for learning and refining a concept, since Patrick Winston's influential work, looks like this: for every new positive instance, build a generalization of the discrimination rule that the new instance will also satisfy; for every new negative instance, build a specialization of the rule that the new instance will not satisfy. In other words, Winston views learning as a heuristic search in a space of symbolic descriptions, driven by an incremental process of specialization and generalization.

Inductive systems include Ryszard Michalski's "conceptual clustering" and Tom Mitchell's "version space". Michalski's method is "data-driven" like Winston's: the symbolic description is built bottom-up from the set of instances.

Mitchell's method is instead "model-driven", which means that symbolic descriptions are predefined and instances select the most appropriate one. In Mitchell's case, all concepts and their abstractions are represented in a space which is partially ordered by the relation of generality. An incremental process of refinement narrows down the space to one description. New instances "shrink" down the space by generalizing the set of minimal elements and specializing the set of maximal elements. As new instances keep shrinking down the space, the concept gets defined more and more accurately until the two sets of minimal and maximal elements are the same set. Mitchell's "version space" seems to be psychologically plausible because concepts are indeed learned and refined over a period of time, their vagueness slowly turning into crispness.

Programs That Learn: Deduction

The analytic paradigm, instead, utilizes past problem solving experience to formulate the search strategy in the space of potential solutions. Deductive learning systems include Paul Rosenbloom's "chunking", Jerry DeJong's "explanation-based learning", Jaime Carbonell's "derivational analogy", John Holland's "classifiers".

Rosenbloom's programs are aimed at simulating the law of practice: the time required to perform an action decreases exponentially with the number of times the action is performed. His "chunking" technique progressively reduces the amount of processing needed to determine what action must be taken in the face of a situation. Ultimately, it tends to reduce every situation-action pair to a stimulus-response pair that does not require any "thinking" at all.

An explanation-based learning system (inspired by Richard Fikes' work) is given a high-level description of the target concept, a single positive instance of the concept, a description of what a concept definition is and domain knowledge. The system generates a proof that the positive instance satisfies the target concept and then generalizes the proof.

Learning by analogy was originally investigated by Patrick Winston, who focused on learning a concept analogous to another concept (which resulted in a transfer of features from a frame to another frame). Carbonell applied the method to sequences of operators rather than to features. Derivational analogy solves a problem by tweaking a plan (represented as a hierarchical goal structure) used to solve a previous problem.

Theory Formation

Learning a concept is actually not a big deal. Many concepts must be learned to perform even the simplest of daily tasks, and once many concepts are learned they must be combined in a "theory" of the domain if they have to make any sense at all. Given a theory of the domain, then an individual or a system can plan meaningful actions in that domain. Theory formation turns out to be quite tricky. A group of concepts can be combined in infinite ways, and most are not very useful. Physics is a good example of a theory: concepts abound, from mass to electricity, but they are held together by just a few laws.

Douglas Lenat believed that theories can be built only by using "rules of thumb" on what a theory is and how it usually looks like. In other words, some concepts are more interesting than others, and some relations between concepts are more interesting than others. Lenat's heuristics plays the role of a scientist's intuition.

Lenat's approach with respect to building theories was model-driven. Pat Langley's approach, instead, was data-driven: given experimental data, build a hierarchy of hypotheses and eventually a full-fledged theory that explains them. The only rule of thumb is that regularity matters and everything else does not: any theory is a theory of the regularities that occur in a domain.

Either way, one needs heuristics (intuition, rules of thumb, common sense) in order to learn a new theory. Both Lenat and Langley got intrigued by the origins of heuristics and started studying how heuristics itself can be learned. In other words: how does one progress from being a novice, who is moving blindly around the environment and is capable only of applying rigid rules, to being an expert, who relies on intuition and rules of thumbs? For Lenat this meant that one had to progress from using weak methods to using domain-specific methods through a process of generate and test (generate a strategy, test it, tweak it, and so forth). Tom Mitchell's approach was similar, but aimed at generating the version space.

All of these are attempts at building machines that can learn. All of them are extremely limited in how and what they can learn.

Notwithstanding these attempts at building knowledge-based programs that can learn, learning has remained a liability, not an asset, of the field, especially when compared with the achievement of neural networks.

Genetic Algorithms

In the 1970s the US computer scientist John Holland had the intuition that the best way to solve a problem is to mimic what biological organisms do to solve their problem of survival: to evolve (through natural selection) and to reproduce (through genetic recombination). Genetic algorithms apply recursively a series of biologically-inspired operators to a population of potential solutions of a given problem. Each application of operators generates new populations of solutions, which should better and better approximate the best solution. What evolves is not the single individual but the population as a whole.

Genetic algorithms are actually a further refinement of search methods within problem spaces. Genetic algorithms improve the search by incorporating the criterion of "competition".

Recalling Newell's and Simon's definition of problem solving as "searching in a problem space", David Goldberg defines genetic algorithms as "search algorithms based on the mechanics of natural selection and natural genetics". Most optimization methods work from a single point in the decision space and employ a transition method to determine the next point. Genetic algorithms, instead, work from an entire "population" of points simultaneously, trying many directions in parallel and employing a combination of several genetically-inspired methods to determine the next population of points.

One can employ simple algorithms such as "reproduction" (that copies chromosomes according to a fitness function), "crossover" (that switches segments of two chromosomes) and "mutation", as well as more complex algorithms such as "dominance" (a genotype-to-phenotype mapping), "diploidy" (pairs of chromosomes), "abeyance" (shielded against over-selection), "inversion" (the primary natural mechanism for recording a problem, by switching two points of a chromosome); and so forth.

Holland's "Classifier" (which learns new rules to optimize its performance) was the first practical application of genetic algorithms. A classifier system is a machine-learning system that learns syntactically rules (or "classifiers") to guide its performance in the environment. A classifier system consists of three main components: a production system, a credit system (such as the "bucket brigade") and a genetic algorithm to generate new rules. Its emphasis on competition and cooperation, on feedback and reinforcement, rather than on pre-programmed rules, set it apart from knowledge-based models of Artificial Intelligence.

A measure function computes how "fit" an individual is. The selection process starts from a random population of individual. For each individual of the population the fitness function provides a numeric value for how much the solution is far from the ideal solution. The probability of selection for that individual is made proportional to its "fitness". On the basis of such fitness values a subset of the population is selected. This subset is allowed to reproduce itself through biologically-inspired operators of crossover, mutation and inversion.

Each individual (each point in the space of solutions) is represented as a string of symbols. Each genetic operator performs an operation on the sequence or content of the symbols.

When a message from the environment matches the antecedent of a rule, the message specified in the consequent of the rule is produced. Other messages produced by the rules cycle back into the classifier system, some generate action on the environment. A message is a string of characters from a specified alphabet. The rules are not written in the Predicate Logic of expert systems, but in a language that lacks descriptive power and is limited to simple conjunctive expressions.

Credit assignment is the process whereby the system evaluates the effectiveness of its rules. The "bucket brigade" algorithm assigns a strength (a measure of its past usefulness) to each rule. Each rule then makes a bid (proportional to its strength and to its relevance to the current situation) and only the highest-bidding rules are allowed to pass their messages on. The strengths of the rules are modified according to an economic analogy: every time a rule bids, its strength is reduced by the value of the bid while the strength of its "suppliers" (the rules that sent the messages matched by this bidder) are increased. The bidder's strength will in turn increase if its consumers (the rules that receive its message) become bidders. This leads to a chain of suppliers/consumers whose success ultimately depends on the success of the rules that act directly on the environment.

Then the system replaces the least useful (weak) rules with newly generated rules that are based on the system's accumulated experience, i.e. by combining selected "building blocks" ("strong" rules) according to some genetic algorithms.

Holland went on to focus on "complex adaptive systems". Such systems are governed by principles of anticipation and feedback. Based on a model of the world, an adaptive system anticipates what is going to happen. Models are improved based on feedback from the environment.

Complex adaptive systems are ubiquitous in nature. They include brains, ecosystems and even economies. They share a number of features: each of these systems is a network of agents acting in parallel and interacting; behavior of the system arises from cooperation and competition among its agents; each of these systems has many levels of organization, with agents at each level serving as building blocks for agents at a higher level; such systems are capable of rearranging their structure based on their experience; they are capable of anticipating the future by means of innate models of the world; new opportunities for new types of agents are continuously being created within the system.

All complex adaptive systems share four properties (aggregation, non-linearity, flowing, diversity) and three mechanisms (categorization by tagging, anticipation through internal models, decomposition in building blocks).

Each adaptive agent can be represented by a framework consisting of a performance system (to describe the system's skills), a credit-assignment algorithm (to reward the fittest rules) and a rule-discovery algorithm (to generate plausible hypotheses).

Emergent Computation

Emergent computation is to sequential computation what nonlinear systems are to linear systems: it deals with systems whose parts interact in a nontrivial way. Both Alan Turing and John Von Neumann, the two mathematicians who inspired the creation of the computer, were precursors in emergent computation: Turing formulated a theory of self-catalytic systems and Von Neumann studied self-replicating automata.

In the 1950s Turing introduced the "Reaction-diffusion Theory" of pattern formation, based on the bifurcation properties of the solutions of differential equations.

Turing devised a model to generate stable patterns:
- X catalyzes itself: X diffuses slowly
- X catalyzes Y: Y diffuses quickly
- Y inhibits X
- Y may or may not catalyze or inhibit itself

Some reactions might be able to create ordered spatial schemes from disordered schemes. The function of genes is purely catalytic: they catalyze the production of new morphogenes, which will catalyze more morphogenes until eventually form emerges.

Von Neumann saw life as a particular class of automata (of programmable machines). Life's main property is the ability to reproduce. In the 1940s Von Neumann had already proven that a machine could be programmed to make a copy of itself.

Von Neumann's automaton was conceived to absorb matter from the environment and process it to build another automaton, including a description of itself. Von Neumann realized (years before the genetic code was discovered) that the machine needed a description of itself in order to reproduce. The description itself would be copied to make a new machine, so that the new machine too could copy itself.

In Von Neumann's simulated world, a large checkerboard was a simplified version of the real world, in which both space and time were discrete. Time, in particular, was made to advance in discrete steps, which meant that change could occur only at each discrete step, and simultaneously for everything that had to change.

Von Neumann's studies of the 1940s led to an entire new field of Mathematics, called "Cellular Automata". Technically speaking, cellular automata are discrete dynamical systems whose behavior is completely specified in terms of a local relation. In practice, cellular automata are the computer scientist's equivalent of the physicist's concept of field. Space is represented by a uniform grid and time advances in discrete steps. Each cell of space contains bits of information. Laws of nature express what operation must be performed on each cell's bits of information, based on its neighbor's bits of information. Laws of nature are local and uniform. The amazing thing is that such simple "organisms" can give rise to very complex structures, and those structures recur periodically, which means that they achieve some kind of stability.

Von Neumann understood the dual genetics of self-reproducing automata: namely, that the genetic code must act as instructions on how to build an organism and as data to be passed on to the offspring. This was basically the idea behind what will be called DNA: DNA encodes the instructions for making all the enzymes and the protein that a cell needs to function and DNA makes a copy of itself every time the cell divides in two. Von Neumann indirectly understood other properties of life: the ability to increase its complexity (an organism can generate organisms that are more complex than itself) and the ability to self-organize.

When a machine (e.g., an assembly line) builds another machine (e.g., an appliance), there occurs a degradation of complexity, whereas the offspring of living organisms are at least as complex as their parents and their complexity increases in evolutionary times. A self-reproducing machine would be a machine that produces another machine of equal or higher complexity.

By representing an organism as a group of contiguous multi-state cells (either empty or containing a component) in a 2-dimensional matrix, Von Neumann proved that a Turing-type machine that can reproduce itself could be simulated by using a 29-state cell component.

Turing proved that there exists a "universal computing machine". Von Neumann proved that there exists a universal computing machine which, given a description of an automaton, will construct a copy of it, and, by extension, that there exists a universal computing machine which, given a description of a universal computing machine, will construct a copy of it, and, by extension, that there exists a universal computing machine which, given a description of itself, will construct a copy of itself.

Artificial Life

Another approach to building intelligent programs is based on "Artificial Life" (a term coined by Chris Langton in 1987). "Intelligence" (or, better, "cognition") cannot do without life. Intelligence is a product of life, and it is an evolutionary product of the evolution of life. On the other hand, there is more and more evidence to support the mirror view: that life is very much about cognition, that all life is "cognitive" in nature.

The first computer viruses were produced at Bell Labs in 1962 (the term was coined by David Gerrold in his novel "When Harley was one"). When computer viruses became famous, they simply popularized the discipline that was attempting to build self-replicating automata at software level, a school of thought started by Von Neumann decades earlier. Self-replication (the ability to produce offspring from self-contained instructions) is the prerequisite to evolution.

It turns out that self-replicating and evolving systems can also replace expert systems.

Artificial Intelligence solves a problem by reasoning about the knowledge of the problem's domain. Artificial Life ("Alife") lets possible solutions "evolve" in that domain until they fit the problem. Sometimes there is no perfect solution, just a "best fit". Solutions evolve in populations according to a set of "genetic" algorithms à la John Holland that mimic biological evolution. Each generation of solutions, as obtained by applying those algorithms to the previous generation, is better "adapted" to the problem at hand.

In 1952 the Norwegian mathematician Nils Barricelli became the first person to actually run artificial evolution experiments on computers (basically, a one-dimensional cellular automaton).

In 1970 the US computer scientist Michael Conrad and the US physicist Howard Pattee developed one of the earliest artificial models of life, that modeled competition among individuals equipped with a genotype representation and a phenotype obtained by interpreting the genotype as instructions.

Likewise, software environments such as the "Tierra" program, developed in 1992 by an US ecologist, Thomas Ray, simulate a world and an evolving population of organisms. Tierra is populated with digital organisms that compete for space in the computer memory and for time in the computer processor. Whatever space and time they manage to get, they use it to reproduce themselves. Like with most simulations of this type, a digital organism's phenotype is also its genotype (the genome is also the body, or viceversa).

Ray draws a distinction between two types of Alife: weak (simulation of life) and strong (synthesis/instantiation of life). The difference is that one is man-made while the other has evolved to be living from inanimate "matter". Tierra, for example, starts out with instances of a simple replicating code and is left to evolve into a living system capable of metabolizing, reproducing and evolving while it interacts with its environment. Ray focuses on "the second major event in the history of life, the origin of diversity."

As Langton points out, the key concept in Alife is "emergent behavior".

These virtual worlds are more than simple simulations of algorithms. They may well be philosophical investigations into the very nature of the universe. For example, the Italian physicist Tommaso Toffoli speculated that the universe could be viewed as a computer. Frank Tipler points out that, at the very least, there is no way to tell a computer simulation of the real world from the real world, as long as one is inside the simulation. A simulated observer would perceive the simulated world exactly the same way that the real observer perceives the real world. Any test to reveal whether her world is the real world would succeed, by definition. Therefore, there is also no way for me to tell whether I am a simulated observer inside a simulated universe, or a real observer inside a real universe. Therefore, the distinction between reality and simulation becomes fictitious.

Most evolutionary engineering is software-based. Hardware-based simulations of natural evolution are based on the idea of a software bit string that is used to configure programmable logic devices as a genetic algorithm chromosome, so that the configuration of the circuit will evolve at electronic speed. The final goal is to build machines that evolve independently, or, more properly, are "evolvable hardware" (a discipline that was officially born in 1995). The Swiss computer scientist Daniel Mange builds electronic circuits that can grow/evolve rather than be designed. Mange's "embryological electronics" employs field programmable gate arrays that exhibit the ability to reproduce the circuit of any programmable function and to self-repair. The "Firefly Machine", for example, is based on a variation of Von Neumann's cellular programming techniques: parallel cellular machines evolve to solve a problem. The "Embryonics" project deals with ontogeny, or growth: just like any multicellular organism grows over its lifetime, so a multicellular automata should exhibit embryonic development driven by the same processes of cellular division and differentiation.

Another center for biologically-inspired systems is the Evolvable Systems Lab in Japan, headed by Tetsuiya Higuchi.

Basically, Artificial Life replaced the "problem solver" of Artificial Intelligence with an evolving population of problem solvers. The "intelligence" required to solve a problem is not in an individual anymore, it is in an entire population and its successive generations; it is not due to the knowledge of a solver, but to the evolutionary algorithms of nature that operate on the genetic code of a population.

It is not the solver who is smart enough to solve the problem, but the knowledge she has. It is evolution that eventually builds the solver who is smart enough to solve the problem using the knowledge that is available.

Life As Virtual Reality

David Deutsch views the technology of virtual reality (the ability of a computer to simulate a world) as the very technology of life. Virtual reality is a "physical embodiment of theories about an environment". So defined, virtual reality is an important property of nature: it is life itself. Genes embody knowledge about their ecological niche. An organism is merely the immediate environment which copies the replicators (the organism's genes). The genes, on the other hand, represent the survival of knowledge, knowledge about the environment. An organism is a virtual-reality rendering of the genes. Therefore, living processes and virtual-reality rendering are the same kind of process. Thus virtual reality is not only a property of computers but a general property of nature, the very essence of life itself.

Holism

Science after the Cartesian revolution has employed a reductionist paradigm. The strongest opposition to the reductionist paradigm came from the Catholic Church, not from scientists. Artificial Intelligence is one of the first scientific disciplines that has been forced to resurrect the holistic methods to analyze problems and systems. The heuristics of knowledge-based systems, the genetic algorithms and the neural networks are all cases in which the whole is not analyzed as the sum of its parts but as a whole.

Reductionism assumes that one can produce a mathematical model of the system, and that one can divide a problem into subproblems in order to reduce its complexity. Artificial Intelligence is, de facto, a philosophical alternative to the program of reductionism There is no model of the system. The holistic methods (trial and error, pattern recognition, etc) are model-free. The US system theorist Kirstie Bellman ("New Architectures for Constructed Complex Systems", 1999) and the US philosopher Stephen Kercel believe that holistic methods are more suitable for "bizarre systems", i.e. systems that do not lend themselves to a mathematical model (e.g., human language, medicine, consciousness).

The Turing Test

In 1950 Turing proposed a test that came to be interpreted as a test to determine whether a machine is intelligent or not: a computer can be said to be intelligent if its answers are indistinguishable from the answers of a human being. The test can be performed on the machine alone or on the machine and a human. In the former case, the "observer" of the test must be led to believe, by the machine's cunning answers, that the tested thing is a human. In the latter case, the observer must be incapable of telling which are the answers of the human and which are the answers of the machine.

Turing's article ("Computing machinery and intelligence") started the quest for the "intelligent" computer that led to Artificial Intelligence.

Regardless of what Artificial Intelligence has achieved so far, a debate has been raging about whether an intelligent machine is possible or not at all. After the invention of the computer, a number of thinkers from various disciplines (Herbert Simon, Allen Newell, Noam Chomsky, Hilary Putnam, Jerry Fodor) adopted a paradigm modeled after the relationship between the hardware and the software of a computer. They basically reduced "thinking" to the execution of an algorithm in the brain.

John Searle is the foremost opponent of Artificial Intelligence. He argues that computers are purely syntactical and therefore cannot be said to be thinking. In his thought experiment of the "Chinese room", a man who does not know how to speak Chinese, but is provided with formal rules on how to build perfectly sensible Chinese answers, would pass the analogous Turing test for understanding Chinese, even if he never knows what those questions and those answers are about. That opened the floodgates of the arguments that computation per se will never lead to intelligence. Searle's Chinese room argument can be summarized as follows: computer programs are syntactical; minds have a semantics; syntax is not by itself sufficient for semantics. Whatever a computer is computing, the computer does not "know" that it is computing it: only a mind can look at it and tell what it is.

Paraphrasing Fred Dretske, a computer does not know what it is doing, therefore "that" is not what it is doing. For example, a computer does not compute 1+1, it simply manipulates symbols that eventually will yield the result of 1+1.

Countless replies have been provided. Some have observed that the man may not "know" Chinese, but the room (i.e., the man plus the rules to speak Chinese) does qualify as a fluent Chinese speaker. Some have found flaws in the premises (the theorem sets to prove what has been stated in the premises, that the man does not understand Chinese).

And, ultimately, it all depends on the definition of the word "understand". In a sense, Searle has simply slowed down and broken down the process of understanding, but what we do when we understand something is precisely what the man does in the room. So Searle's objection is simply about the size of the information and the speed of information processing, and we would all assume that the man understands Chinese if he performed his task in a few milliseconds with the help of miniaturized microfilms invisible to us. Searle's objection sounds more like: if you can tell what the mechanism is that produces "understanding", then that cannot be true "understanding".

Searle does concede that a brain is a machine and that, in principle, we could build a totally equivalent machine that would then have consciousness. He does not agree that a computer is such a machine.

Computation as defined by Turing is not sufficient to grant the presence of thinking.

The simulation of a mind is not itself a mind.

Experience vs Knowledge

Inspired by Edmund Husserl's phenomenology, another US philosopher and critic of Artificial Intelligence, Hubert Dreyfus, thinks that comprehension can never do without the context in which it occurs. The information in the environment is fundamental for a being's intelligence.

Dreyfus criticized the four fundamental assumptions of Artificial Intelligence: biological (that the brain must operate as a symbolic processor), psychological (that the mind must obey a heuristic program), epistemological (that there must be a theory of practical activity) and ontological (that the data necessary for intelligent behavior must be discrete, explicit and determinate). In his opinion all of them are just not plausible. Furthermore, Dreyfus emphasizes the role of the body in intelligent behavior which knowledge-based systems neglect. Human experience is intelligible only insofar as it gets organized in terms of a situation (as a function of human needs).

Dreyfus presents a model of acquisition of performance by humans structured in five stages. First, we are born novices: we simply follow the rules (an instructor, a manual). The moves of novices are not secure and not fluid, although they can be technically correct. Sometimes applying a rule is plain silly, but the novice will still do so because he doesn't know better. Then we become advanced beginners. At this stage we are capable of modifying rules based on the situation. Our behavior is still driven by rules but it doesn't look as awkward. Competent humans, the next stage, follow rules but in a very fluid manner, and their rules are much more plastic: the competent human knows that she can modify the rules, and she will feel guilty if something goes wrong, even if she followed the proper rules. Proficient performers do not even follow rules anymore: they act by reflex. The fact that they have encountered similar situations many times matters more than the original rules. Experts, the final stage, do not even remember the rules. Sometimes if they have to articulate them they can't even figure them out. They just act based on their expertise and their intuition. They are often not even aware of what they are doing. An expert driver does not realize that she is shifting gears and at which point she is shifting gears. She just shifts gear when it's appropriate to.

Dreyfus points out that a failure usually results in degradation: you step back to a lower stage to understand what went wrong. An expert does not even remember the rules, but, if she can't start the car, she gradually walks down the ladder from expert to merely competent all the way down to novice and will finally pick up the driver's manual.

Only novices behave like expert systems. Human experts behave in a radically different way. An expert has synthesized experience in an unconscious behavior that reacts instantaneously to a complex situation. What the expert knows cannot be decomposed in rules or any other type of discrete knowledge representation; therefore it cannot be emulated by an expert system.

The foundation of Dreyfus's argument is that minds do not use a theory about the everyday world; and the reason is that there is no set of "context-free" primitives of understanding. Human knowledge is skilled "know-how", as opposed to the logical representations that expert systems have to rely upon, or "know-that".

Also drawing from Martin Heidegger's phenomenology, the US computer scientist Terry Winograd is skeptic that intelligence can be due to processes of the type of production systems, i.e. to the systematic manipulation of representations. Intelligent systems act, don't think. People are "thrown" in the real world and cannot afford to deal with all the possible alternatives of a situation. They think only when action does not yield the desired result. Only then do they pause to picture the situation in its complexity and decompose it into its constituents, and try to infer action from knowledge. But, again, this behavior is more typical of the novice than of the expert.

Another way to see the same argument is to consider what makes an expert so much more efficient at solving a problem: the first few seconds. A chess champion wins the game against a novice because of the first few moves, not because of the huge knowledge that the champion has and could use against the novice. That huge knowledge is, in turn, certainly important to determine the first moves (that will deliver a crippling blow to the novice).

The US computer scientist Rodney Brooks ("A robust layered control system for a mobile robot", 1986) offered an alternative way of achieving Artificial Intelligence, which significantly revised the foundations of the symbol-processing program: he argues that intelligence cannot be separated from the body. Intelligence is not only a process of the brain, it is embodied in the physical world. Every part of the body is performing an action that contributes to the overall "functioning" of the organism in the environment. There is no need for a central representation of the world, so long as all component tasks help each other operate in the world. Cognition is grounded in the physical interactions with the world. Intelligence "is" about moving in a physical world and cannot exist without a physical world.

Dreyfus and Winograd created the schism between "Cartesian" and "Heideggerian" Artificial Intelligence programs. A requirement of the latter is the ability to work in "thrown" situations without building internal representations. Philip Agre built the first "Heideggerian AI", Pengi, a system that played the arcade videogame Pengo.

Goedel's Limit

Following the British philosopher John Randolph Lucas ("Minds, Machines and Gödel", 1961), the British physicist Roger Penrose resorted to Goedel's theorem to undermine the very foundations of Artificial Intelligence. Goedel's theorem states that every formal system (which is bigger than Arithmetic) contains a statement that cannot be proven true or false. Indirectly, Goedel's theorem states the preeminence of the human mind over the machine: some mathematical operations are not computable, nonetheless the human mind can treat them (at least to prove that they are not computable). Humans can realize what Goedel's theorem states, whereas a machine, limited to mathematical reasoning, would never realize what it states. We can intuitively comprehend a truth that the computer can only try (and, in this case, fail) to prove. Therefore a computer will never be equal to a mind. And, in general, no mathematical system can fully express the way my mind thinks.

Again, countless replies, have been provided.

First of all (Hilary Putnam), a computer can observe the failure of "another" computer's formal system, just like a human mind can observe it. A computer can easily prove the proposition "if the theory is consistent, then the proposition that there is at least one undecidable proposition is true". Which is exactly all the human mind is capable of doing.

Second, even if Goedel's theorem sets a limit, it is not a limit of the machine, it is a limit of the human mind: the human mind will never be capable of building a machine that can think. This does not prove that machines cannot think.

Third, Penrose's demonstration can be used to prove that a machine cannot prove the validity of a mathematical demonstration, a fact that is contradicted by our experience.

Fourth, Goedel's theorem is false in some nonstandard mathematical systems, as, for example the Zimbabwe-born mathematician Aaron Sloman pointed out ("The Emperor's Real Mind", 1992). One of Goedel's conditions is that the mathematical system must be consistent (i.e., not contain a contradiction), but that can only be if the undecidable statement is added to the system, assuming either true or false. Nonstandard models assume that it is false. Goedel's theorem, because of the way Goedel carried it out (by employing infinite sets of formulas), leaves the illusion of proving a truth which in reality is never proved, cannot be proven and must be arbitrarily decided.

Fifth, Rudy Rucker believes that conscious machines could be built, following an observation of Goedel himself, that we cannot build a machine that has our mathematical intuition but such a machine can exist and can be discovered by humans. If such a machine exists, humans cannot understand its functioning. Such a machine cannot be built by humans, but could be built by Darwinian evolutionary steps starting from a man-made machine. If a machine can be built that exhibits a behavior completely similar to that of humans, then a machine can be built that is as conscious as humans. What Goedel's theorem asserts is that "the human mind is not capable of formulating all of its mathematical intuitions" (quoting Goedel himself).

Sixth, the British physicist Stephen Hawking notes that the behavior of earthworms can probably be simulated adequately with a computer, because they do not worry about Goedel sentences. Darwinian evolution can generate human intelligence from earthworm intelligence through a process (natural selection) for which Goedel's theorem is also irrelevant. Therefore, Goedel's theorem does not forbid the birth of an intelligent computer.

What Goedel's theorem proves, if anything, is an intrinsic limit to "any" form of intelligence, including the machine's but also Penrose's...

The Turing Test Revisited

The Turing Test has been credited with starting a whole new branch of science. That is surprising, given that the test itself is not formulated in a scientific manner at all.

First of all, it is not clear whether Turing was concerned with intelligence, mind or consciousness. Is his test supposed to reveal whether a machine is intelligent or cognitive or conscious? The three are quite different. Nowhere does Turing bother to distinguish among them, though. Intelligence comes in degrees. Animals are intelligent, to some degree. It is debatable whether they are capable of thinking (conscious). A mentally-retarded person may not be intelligent, but she is probably conscious. Turing does not discriminate and therefore does not tell us what his test is supposed to measure.

Second, when one proposes a test to the scientific community, one must be specific about the setting (e.g., what instruments will be used). Turing's test uses a human being to decide whether a machine is as good as another human being. Thus both the instrument and one of the quantities to be measured are humans: good scientific policy would have required him to be specific about both. He does not provide a definition or a prescription for what the observer must be. Can a mentally retarded person perform the test? Can somebody under the influence of drugs perform it? Or does it have to be the most intelligent human? (The result of the test will obviously vary wildly depending on which one Turing chooses). As for the human to be tested against the machine, Turing doesn't specify which type of human he wants to test: a priest, an attorney, an Australian aborigine, an avid reader of porno magazines, a librarian, a physician, an economist...?

The observer has to determine whether the answers to her questions come from a human or a machine. If the unknown "answerer" is the machine, and she is led to think that it is a human, then the machine qualifies as "intelligent" (or cognitive or conscious, we are not sure). But Turing does not tell us what conclusions we have to draw if the unknown answerer is a human and the observer is led to think from her answers that she is a machine. In other words, if a machine fails the test, then Turing concludes that it is not intelligent: but what does Turing conclude if a human fails the test? That humans are not intelligent?

Therefore, one of the reasons why Turing's paper has led to so much controversy is that Turing was not clear enough about what he was saying.

Can a machine be a cognitive system? If one circumscribes cognition to the processes of remembering, learning and reasoning, most scientists would agree that, yes it is possible. Does that make it "intelligent": in the sense that most people use that word, yes, probably yes. Does it make it also aware of being what it is? Not necessarily so. These are different questions for which there probably are different answers. But, if it were a well-formulated question, most people would agree on the answer.

Even if it were restated in a scientific manner, the Turing test per se would probably not amount to much: it is not right or wrong, but simply meaningless. Even if a machine could answer all questions, what would that prove? If we found a "thing" that can answer all our questions but does not eat, move, feel emotions and so forth, we would just consider it as a very sophisticated machine, not a human being. That is what the Turing test measures: how good the machine is at answering questions, nothing more.

The thesis of Turing's test can be restated as: "Can a machine be built that will fool a human being into believing it is another human being?" Nowhere in his writings did Turing prove the equivalence of this question with the question "Can a machine think?" If we answer "yes" to the first question, we don't necessarily answer "yes" to the second.

As the US computer scientist Stuart Russell remarked, Turing's definition is at the same time too weak and too strong. Too weak because it does not include "intelligent" behavior such as "dodging bullets" and too strong because it does include unintelligent beings such as Searle's

Chinese-room translator. Most children, who cannot answer a lot of questions that an adult could answer, would not pass the test, but that does not make them machines. Turing's is a partial extensional definition, that fails to capture the intensional definition of intelligence.

Would we consider our peer an object or being that is not alive? To be conscious without being alive is a piece of nonsense. Before we ask whether machines can think, we should therefore ask whether they can be alive.

Biological systems undergo growth. Machines cannot undergo physical growth. In machines only the "mind" can grow over time. In biological systems the "mind" grows with the rest of the body. In machines the "mind" may never decay. In biological systems the mind decays with the rest of the body. The mind is closely tied to the body. For most people a mind without a body (that grows with it) is just not a mind.

Machines and Real Life

The very reason that people started wondering if computers could ever think is that computers are very fast and have huge memories. But do they? One could argue that the computer has no memory at all: it remembers what I want to remember. What we call "the computer's memory" is in reality just an extension of my memory. I remember what I was doing five minutes ago and five hours ago and, if I focus, maybe I can remember what I was doing five months ago. But the computer has no "memory" of what it was doing five seconds ago. What we call "memory" in the case of a computer is something completely different from what we call "memory" in the case of animals. In a sense, the whole issue arises only because someone decided to use that word, "memory", for that component of my computer that stores data (my data) and instructions (the application's instructions). If we had called it "heart" maybe today philosophers would be discussing if the computer has a heart. Ditto for speed. The computer is "fast" at something that we don't call "fast". We say that a chef was fast if she cooked the meal in record time: is the computer fast at cooking a meal in the kitchen? The fastest robot in the world takes forever just to extend a hand in the right direction. What we call "speed" has to do with crossing streets, planting tomatoes, dusting shelves and walking up and down the stairs. The computer is actually extremely slow at any of these. It is in fact slower than any animal that ever existed.

Again, it is just syntax: we called them "speed" and "memory" to reuse existing words but they are neither speed nor memory.

Machine Charisma

Nonetheless, it is important to realize why we are interested in knowing whether a computer can think, but not whether a refrigerator can think. They are both complex machines, and it is not obvious which one is more indispensable. In the event of a catastrophic earthquake, most people will be more concerned about the food in their refrigerator than about the files in their computer. Computers only solve some problems, not all problems: they don't wash dishes, don't run on roads, and don't make ice. Ordinary life (and certainly survival) is more than mere Math.

One reason for our fixation with computers is purely socio-historical. In the age when Artificial Intelligence was born, computers were huge, and the sheer size was commanding attention. No other machine was that big (and nobody could predict that they would become so small so quickly). Because they were so terribly primitive (and this borders on the paradoxical), they were also very difficult to operate, and it was commonly held that only super-intelligent humans (and many of them together) were able to control them. The (flawed) syllogism was that if they required so much intelligence to operate, then computers must be very intelligent.

Furthermore, for anybody who was not fluent in electronics it was difficult to conceive of computation without thought: animals can do many things, but not multiplication, and certainly not faster than humans. Math was a privilege of one higher primate: us. It was psychologically easier to assume that computers could think than to tear down a thousand-year old habit of associating Mathematics with the ability to think. Finally, a lot of the excitement arose simply from confusing intelligence, cognition and consciousness, a mistake that in the age before the boom of neurophysiology was very common, even among the likes of Turing.

The one thing that computers taught us is precisely that our traditional views of intelligence were flawed and we needed better ones. Before computers were invented, the scientific community had never been forced to define and distinguish intelligence, cognition and consciousness. No animal was threatening human superiority in any of them. Computers forced us to do that.

Even if intelligence is just computing many numbers very quickly (which means that a palmtop computer is intelligent), that does not automatically entail cognition or consciousness. And viceversa: a mentally retarded person may not be able to perform a multiplication but still be capable of feelings and introspection.

Computers have convinced us that "intelligence" is simply a misconception, and the word is not scientific.

The Turing Test was a misconception born out of a misconception.

What is Intelligence?

What is "intelligence"? The Turing test was based on the assumption by Newell and Simon that intelligence is about solving problems. Yet, solving problems is nothing special: finding problems is far more difficult. Given enough time and resources, most people would solve any problem. But very few people could come up with the problem in the first place.

For example, very few people wonder why it gets colder as you climb up a mountain. After all, you are moving closer to the sun, which is the main source of heat. It should obviously get warmer as we get closer to the sun. Very few people ever wonder.

Once they are told, they will eventually find the solution to the paradox. All that one has to do is to think about it, consult a book or two, call up a friend. As the philosophers of Artificial Intelligence correctly said, the solution is out there and it is just a matter of "searching" for it.

Once somebody formulates a problem, most people can solve it. We use professional "problem solvers" called "scientists" because we are normally busy doing other things.

The real intelligence is in formulating the problem, in realizing that something we take for granted is not explained by our knowledge.

Artificial Intelligence initially missed the point: somebody who can answer all the questions is not very intelligent, it just has nothing better to do. It resembles a machine more than a human.

The Turing test is not about how human a machine is, but how mechanical a human is. The Turing test tests a human, not a machine.

We can certainly build a machine that will answer all questions. But that has little to do with our "intelligence". That only has to do with "symbolic processing". The real intelligence test would be: can the machine "ask" the questions? Can we build a machine that will ask all the questions that an intelligent human would ask in a given situation? Can a machine "wonder"? Can a machine be the one asking questions and rating the answers in Turing's test?

Further Reading

Agre, Philip: THE DYNAMIC STRUCTURE OF EVERYDAY LIFE (MIT, 1988)

Arbib, Michael: METAPHORICAL BRAIN (Wiley, 1972)

Arbib, Michael: METAPHORICAL BRAIN 2 (Wiley, 1989)

Arbib, Michael: BRAINS MACHINES AND MATHEMATICS (Springer Verlag, 1987)

Ashby, Ross: DESIGN FOR A BRAIN (John Wiley, 1952)

Barr, Avron & Feigenbaum Edward: HANDBOOK OF ARTIFICIAL INTELLIGENCE (William Kaufmann, 1982)

Bernstein, Nicholas: GENERAL BIOMECHANICS (1926)

Boden, Margaret: PHILOSOPHY OF ARTIFICIAL INTELLIGENCE (Oxford, 1990)
Brillouin, Leon: SCIENCE AND INFORMATION THEORY (Academic Press, 1956)
Brooks, Rodney & Luc Steels: THE ARTIFICIAL LIFE ROUTE TO ARTIFICIAL INTELLIGENCE (Lawrence Erlbaum, 1995)
Cannon, Walter: THE WISDOM OF THE BODY (Norton, 1932)
Carbonell, Jaime: MACHINE LEARNING (MIT Press, 1989)
Charniak, Eugene: ARTIFICIAL INTELLIGENCE PROGRAMMING (Lawrence Erlbaum, 1987)
Cohen, Fred: IT'S ALIVE (Wiley, 1994)
Cziko Gary: THE THINGS WE DO (MIT Press, 2000)
Deutsch, David: THE FABRIC OF REALITY (Penguin, 1997)
Dreyfus, Hubert: WHAT COMPUTERS CAN'T DO (Harper & Row, 1979)
Dreyfus, Hubert & Dreyfus Stuart: MIND OVER MACHINE (Free Press, 1985)
Feigenbaum, Edward: COMPUTERS AND THOUGHT (MIT Press, 1995)
Frost, Richard: INTRODUCTION TO KNOWLEDGE BASED SYSTEMS (MacMillan, 1986)
Genesereth, Michael & Nilsson Nils: LOGICAL FOUNDATIONS OF ARTIFICIAL INTELLIGENCE (Morgan Kaufman, 1987)
Graubard, Stephen: THE ARTIFICIAL INTELLIGENCE DEBATE (MIT Press, 1988)
Hofstadter, Douglas: GOEDEL ESCHER BACH (Vintage, 1980)
Holland, John: ADAPTATION IN NATURAL AND ARTIFICIAL SYSTEMS (Univ of Michigan Press, 1975)
Holland, John: EMERGENCE (Basic, 1998)
Holland, John: HIDDEN ORDER (Addison Wesley, 1995)
Koza, John: GENETIC PROGRAMMING (MIT Press, 1992)
Langton, Christopher: ARTIFICIAL LIFE (MIT Press, 1995)
Levy, Steven: ARTIFICIAL LIFE (Pantheon, 1992)
Li, Ming & Vitanyi Paul: AN INTRODUCTION TO KOLMOGOROV COMPLEXITY (Springer-Verlag, 1993)
Luger, George: COMPUTATION AND INTELLIGENCE (MIT Press, 1995)
Maes, Patti: DESIGNING AUTONOMOUS AGENTS (MIT Press, 1990)
Michalski, Ryszard, Carbonell Jaime & Mitchell Tom: MACHINE LEARNING I (Morgan Kaufman, 1983)

Michalski, Ryszard, Carbonell Jaime & Mitchell Tom: MACHINE LEARNING II (Morgan Kaufman, 1986)

Newell, Allen & Simon Herbert: HUMAN PROBLEM SOLVING (Prentice-Hall, 1972)

Nilsson, Nils: THE MATHEMATICAL FOUNDATIONS OF LEARNING MACHINES (Morgan Kaufmann, 1990)

Nilsson, Nils: PRINCIPLES OF ARTIFICIAL INTELLIGENCE (Tioga, 1980)

Pearl, Judea: HEURISTICS (Addison Wesley, 1984)

Powers, William: BEHAVIOR: THE CONTROL OF PERCEPTION (Aldine, 1973)

Ray, Thomas: ZEN AND THE ART OF CREATING LIFE (1994)

Ricoeur, Paul: INTERPRETATION THEORY (1976)

Rucker, Rudy: INFINITY AND THE MIND (Birkhauser, 1982)

Russell, Stuart Jonathan: THE USE OF KNOWLEDGE IN ANALOGY AND INDUCTION (Pitnam, 1989)

Russell, Stuart Jonathan & Norvig Peter: ARTIFICIAL INTELLIGENCE (Prentice Hall, 1995)

Searle John: THE REDISCOVERY OF THE MIND (MIT Press, 1992)

Shannon, Claude & Weaver Warren: THE MATHEMATICAL THEORY OF COMMUNICATION (Univ of Illinois Press, 1963)

Shapiro, Stuart Charles: ENCYCLOPEDIA OF ARTIFICIAL INTELLIGENCE (John Wiley, 1992)

Simon, Herbert: MODELS OF THOUGHT (Yale University Press, 1979)

Simon, Herbert: THE SCIENCES OF THE ARTIFICIAL (MIT Press, 1969)

Sowa, John: KNOWLEDGE REPRESENTATION (Brooks Cole, 2000)

Tarski, Alfred: LOGIC, SEMANTICS, METAMATHEMATICS (Clarendon, 1956)

Turing, Alan: PURE MATHEMATICS (Elsevier Science, 1992)

Turing, Alan: MECHANICAL INTELLIGENCE (Elsevier Science, 1992)

Turing, Alan: MORPHOGENESIS (North-Holland, 1992)

Von Neumann, John: THE COMPUTER AND THE BRAIN (Yale Univ Press, 1958)

Von Neumann, John: THEORY OF SELF-REPRODUCING AUTOMATA (Princeton Univ Press, 1947)

Wolfram, Stephen: A NEW KIND OF SCIENCE (Wolfram, 2002)

Wiener, Norbert: CYBERNETICS (John Wiley, 1948)

Wiener, Norbert: HUMAN USE OF HUMAN BEINGS (1950)

Winograd, Terry & Flores Fernando: UNDERSTANDING COMPUTERS AND COGNITION (Ablex, 1986)

COMMON SENSE: ENGINEERING THE MIND

The Sense of the Mind

In our quest for the ultimate nature of the mind, we are confounded by the very way the mind works. The more we study it, the less it resembles a mathematical genius. On the contrary, it appears that the logic employed by the mind when it has to solve a real problem in a real situation is a very primitive logic, one that we refer to as "common sense", very different from the austere formulas of Mathematics but quite effective for the purposes of surviving in this world. If the mind was shaped by the world, then the way the mind reasons about the world is a clue to where it came from and how it works.

In emergency situations, our rational thinking is often powerless. Common sense determines what we do, regardless of what we think. The puzzling aspect of common sense is that it is sometimes wrong. There are plenty of examples in the history of science of "paradoxes" about common-sense reasoning. Using common-sense reasoning, the Greek philosopher Zeno proved that Achilles could never overtake a turtle. Using common sense reasoning, one can easily prove that General Relativity is absurd (a twin that gets younger just by traveling very far is certainly a paradox for common sense). Common sense told us that the Earth is flat and at the center of the world. Physics was grounded on Mathematics and not on common sense precisely because common sense is so often wrong.

There are many situations in which we teach ourselves to stay "calm", to avoid reacting impulsively, to use our brain. These are all situations in which we know our common sense would lead us to courses of actions that we would probably regret.

Why don't our brains simply use mathematical logic in all their decisions? Why does our common sense tell us things that are wrong? Why can't we often resist the power of that falsehood? Where does common sense come from, and where does its power come from?

Illogical Reasoning

Common sense is a key factor for acting in the real world. We rarely employ classical Logic to determine how to act in a new situation. More often, the new situation "calls" for some obvious reaction, which stems purely from common sense. If we used Logic, and only Logic, in our daily lives, we would probably be able to carry out only a few actions a day. Logic is too cumbersome, and allows us to reach a conclusion only when a problem is "well" formulated. In more than one way, common sense helps us deal with the complexity of the real world. Common sense provides a shortcut to making critical decisions very quickly.

Common sense encompasses both reasoning methods and knowledge that are obvious to humans but that are quite distinct from the tools of classical Logic. When scientists try to formalize common sense, or when they research how to endow a machine (such as the computer) with common sense, they are faced with the limitations of classical Logic. It is extremely difficult, if not utterly impossible, to build a mathematical model for some of the simplest decisions we make. Common sense knows how to draw conclusions even in the face of incomplete or unreliable information. Common sense knows how to deal with imprecise quantities, such as "many", "red", "almost". Common sense knows how to deal with a problem that is so complex it cannot even be specified (even cooking a meal theoretically involves an infinite number of choices). Common sense knows how to revise beliefs based on facts that all of a sudden are proved false. Logic was not built for any of these scenarios.

Furthermore, common sense does not have to deal with logical paradoxes. Paradoxes arising from self-referentiality (such as the liar's paradox) have plagued Logic since the beginning.

A program to ground common sense in Predicate Logic is apparently contradictory, or at least a historical paradox. Science was born out of the need to remove the erroneous beliefs of common sense: e.g., the Earth is not the center of the universe. Science checks our senses and provides us with mathematical tools to figure out the correct description of the world notwithstanding our sense's misleading perceptions. Science was born out of the need to get rid of common sense.

What was neglected is that common sense makes evolutionary sense. Its purpose is not to provide exact knowledge: its purpose is to help an individual survive.

The Demise of Deduction

Logic is based on deduction, a method of exact inference. Its main advantage is that its conclusions are exact. That is the reason why we use it to build bridges or airplane wings. But deduction is not the only type of inference we know. We are very familiar with "induction", which infers generalizations from a set of events, and with "abduction", which infers plausible causes of an effect. Induction has been used by any scientist who has developed a scientific theory from her experiments. Abduction is used by any doctor when she examines a patient. They are both far from being exact, so much so that many scientific theories have been proved wrong over the centuries and so much so that doctors make frequent and sometimes fatal mistakes.

The power of deduction is that no mistake is possible (if you follow the rules correctly). The power of induction and abduction is that they are useful: no scientific theory can be deducted, and no disease can be deducted. If we only employed deduction, we would have no scientific disciplines and no cures.

Alas, deduction works only in very favorable circumstances: when all relevant information is available, when there are no contradictions and no ambiguities. Information must be complete, precise and consistent. In practice, this is seldom the case. The information a doctor can count on, for example, is mostly incomplete and vague. The reason we can survive in a world that is mostly made of incomplete, inexact and inconsistent information is that our brain does not employ deduction in everyday life.

Intuitionism

The limits and inadequacies of classical Logic have been known for decades and numerous alternatives or improvements have been proposed. There are two main approaches: one criticizes the very concept of "truth", while the other simply extends Logic by considering more than two truth values.

As an example of the first kind, "Intuitionism" (a school of thought started in 1925 by the Dutch mathematician Luitzen Brouwer) prescribes that all proofs of theorems must be constructive. Unlike classical Logic, in which the proof of a theorem is only based on rules of inference, in Intuitionistic Logic only "constructable" objects are legitimate. Classical Logic exhibits properties that are at least bizarre. For example, the logical OR operation yields "true" if at least one of the two terms is true; but this means that the proposition "my name is Piero Scaruffi or 1=2" is to be considered true, even if intuitively there is something false in it. Because of this rule, the logical implication between two terms can yield even more bizarre outcomes. A logical implication can be reduced to an OR operation between the negation of the first terms and the second term. The sentence "if x is a bird then x flies" is logically equivalent to "NOT (x is a bird) OR (x flies)". The two sentences yield the same truth values (they are both true or false at the same time). The problem is that the sentence "if the week has eight days then today is Tuesday" is to be considered true because the first term ("the week has eight days") is false, therefore its negation is true, therefore its OR with the second term is true. By the same token, the sentence "Every unicorn is an eagle" is to be considered true (because unicorns do not exist, a fact that makes that formula true).

On the contrary, intuitionists accept formulas only as assertions that can be built mentally. For example, the negation of a true fact is not admissible. Since classical Logic often proves theorems by proving that the opposite of the theorem is false (an operation which is highly illegal in Intuitionistic Logic), some theorems of classical Logic are not theorems anymore.

Intuitionists argue that the meaning of a statement resides not in its truth conditions but in the means of proof or verification.

The "Theory of Types" introduced by the Swedish mathematician Per Martin-Lof in 1970 is an indirect consequence of this approach to demonstration. A "type" is the set of all propositions which are demonstrations of a theorem. Any element of a type can be interpreted as a computer program that can solve the problem represented (or "specified") by the type. This formalizes the obvious connection between Intuitionistic Logic and computer programs, whose task is precisely to "build" proofs.

Alan Gupta's "revisionist theory of truth" also highlights how difficult it is to pin down what "true" really means. Truth is actually impossible to define: in order to determine all the sentences of a language that are true when that language includes a truth predicate (a predicate that refers to truth), one needs to determine whether that predicate is true, which in turn requires one to know what the extension of true is, while such extension is precisely the goal. The solution is to assume an initial extension of "true" and then gradually refine it. Gupta suggests that truth can only be refined step by step. An indirect, but not negligible, advantage of Gupta's approach is that truth becomes a circular concept: therefore all paradoxes that arise from circular reasoning in classical Logic fall into normality.

Frederick and Barbara Hayes-Roth's form of opportunistic reasoning (the "blackboard model" of 1985) stems from the same principles, albeit in a computational scenario. Reasoning is viewed as a cooperative process carried out by a community of agents, each specialized in processing a type of knowledge. Each agent communicates the outcome of its inferential process to the other agents and all agents can use that information to continue their inferential process. Each agent contributes a little bit of truth, that other agents can build on. Truth is built in an incremental and opportunistic manner. Searching for truth is reduced to matching actions: the set of actions the community wants to perform (necessary actions) and the set of actions the community can perform (possible actions). An agent adds a necessary action whenever it runs out of knowledge and has to stop. An agent adds a possible action whenever new knowledge enables it. When an action is made possible that is also in the list of the necessary actions, all the agents that were waiting for it resume their processing. The search for a solution is efficient and more natural, because the only actions undertaken are those that are both possible and necessary. Furthermore, opportunistic reasoning can deal with an evolving situation, unlike classical Logic that considers the world as static.

Classical Logic only admits two "values": true or false. Either a proposition or its negation are true (the "law of the excluded middle"). In 1920 the Polish mathematician Jan Łukasiewicz worked out a logic based on more than just two values. First he added "possible" to "true" and "false". Then he extended the idea to any number of truth values. A logic with more than "true" and "false" is not as "exact" as classical Logic, but it has a higher expressive power. It can be used to better mirror the human experience.

Plausible Reasoning

In our daily lives, we are rarely faced with the task of finding the perfect solution to a problem. If we are running out of gasoline in the middle of the night, we are happy with finding a gas station along our route, even if its gasoline may not be the best or the cheapest. We almost never pause to figure out the best option among the ones that are available. We pick one that leads to a desired outcome. What our mind is looking for all the time is "plausible" solutions to problems, as opposed to "exact" ones. Mathematics demands exact solutions, but in our daily lives we content ourselves with plausible ones. The reason is that sometimes a plausible solution enables us to survive, whereas looking for an exact one would jeopardize our lives. A gazelle who paused to work out the best escape route while a lion is closing in on her wouldn't stand a chance. Often, finding the perfect solution is simply pointless, because by the time we find it the problem would have escalated, i.e. we would be dead.

Classical Logic is very powerful, but lacks this basic attribute: quick, efficient response to problems when an exact solution is not necessary (and sometimes counterproductive).

Several techniques have been proposed for augmenting Logic with "plausible" reasoning: degrees of belief, default rules, inference in the face of absence of information, inference about vague quantities, analogical reasoning, etc.

The Impossibility of Reasoning

A very powerful argument in favor of common sense is that logical reasoning alone would be utterly impossible.

Classical Logic deduces all that is possible from all that is available, but in the real world the amount of information that is available is infinite: the domain must be somehow artificially "closed" to be able to do any reasoning at all. And this can be achieved in a number of ways: the "closed-world assumption" (all relations relevant to the problem are mentioned in the problem statement), "circumscription" (which extends the closed-world assumption to "non-ground" formulas as well, i.e. assumes that as few objects as possible have a given property), "default" theory (all members of a class have all the properties characteristic of the class if it is not otherwise specified). For example, a form of default theory allows us to make use of notions such as "birds fly" in our daily lives. It is obviously not true that all birds fly (think of penguins), but that statement is still very useful for practical purposes. And, in a sense, it is true, even if, in an absolute sense, it is not true. It is "plausible" to claim that birds fly (unless they are penguins).

At the same time, common sense reasoning introduces new problems in the realm of Logic. For example, John McCarthy's "frame problem": it is not possible to represent what does "not" change in the universe as a result of an action, because there is always an infinite set of things that do not change. What is really important to know about the new state of the universe, after an action has been performed? Most likely, the position of the stars has not changed, my name has not changed, the color of my socks has not changed, Italy's borders have not changed, etc. Nevertheless, any reasoning system, including our mind, must know what has changed before it can calculate the next move. A reasoning system must continuously update its model of the world, but McCarthy suggests that this is an impossible task: how does our mind manage?

Complementary paradoxes are the "ramification problem" (infinite things change, because one can go into greater and greater detail of description) and the "qualification problem" (the number of preconditions to the execution of any action is also infinite, as the number of things that can go wrong is infinite). Somehow we are only interested in things that change and that can affect future actions (not just all things that change) and in things that are likely to go wrong (not just all things that can go wrong).

"Circumscription" (McCarthy's solution to the frame problem) deals with default inference by minimizing abnormality: an axiom that states what is abnormal is added to the theory of what is. This reads as: the objects that can be shown to have a certain property, from what is known of the world, are all the objects that satisfy that property. Or: the only individuals for which that property holds are those individuals for which it must hold. (This definition involves a second-order quantifier. Technically, this is analogous to Frege's method of forming the second-order definition of a set of axioms: such a definition allows both the derivation of the original recursive axioms and an induction scheme stating that nothing else satisfies those axioms).

For similar reasons Raymond Reiter introduced the "closed-world axiom" (what is not true is false), or "negation as failure to derive": if a formula cannot be proven using the premises, then assume the formula's negation. In other words, everything that cannot be proven to be true must be assumed to be false. His "Default Logic" employs the following inference rule: "if A is true and it is consistent that B is true, then assume that B is also true" (or "if a premise is true, then the consequence is also true unless a condition contradicts what is known").

Ultimately, these are all tricks to account for how the mind can do any reasoning at all in the face of the gigantic complexity that surrounds it.

Second Thoughts

There is at least one more requirement for "plausible" reasoning. Classical logic is "monotonic": assertions cannot be retracted without compromising the entire system of beliefs. Once something has been proven to be true, it will be true forever. Classical Logic was not designed to deal with "news". But our daily lives are full of events that force us to reexamine our beliefs all the time: our daily system of logic is "nonmonotonic". Therefore, a crucial tool for plausible reasoning is nonmonotonic logic, which allows inferences to be made provisionally and, if necessary, withdrawn at any time. A handful of such logics became popular during the 1980s.

Drew McDermott's formulation of Modal Logic is based on a coherence operator: "P is coherent with what is known" if P cannot be proven false by what is known ("Nonmonotonic Intelligence", 1980).

Robert Moore's "Autoepistemic Logic" ("Semantic Considerations On Nonmonotonic Logic", 1983) is based on the notion of belief (related to McDermott's coherence) and models the beliefs of an agent reflecting upon his own beliefs.

And so forth.

Matthew Ginsberg classified formal approaches to nonmonotonic inference into: proof-based approaches (Raymond Reiter's logic), modal approaches (McDermott's logic, Moore's logic) and minimization approaches (circumscription). Ginsberg argued that a variety of approaches to nonmonotonic reasoning can be unified by resorting to multi-valued logics (logics that deal with more than just true and false statements).

Uncertainty

Another aspect of common sense reasoning that cannot be removed from our behavior without endangering our species is the ability (and even preference) for dealing with uncertainties. Pick any sentence that you utter at work, with friends or at home and it is likely that you will find some kind of "uncertainty" in the quantities you were dealing with. Sometimes uncertainty is explicit, as in "maybe I will go shopping" or "I almost won the game" or "I think that Italy will win the next World Cup". Sometimes it is hidden in the nature of things, as in "it is raining" (can a light shower be considered as "rain"?), or as in "this cherry is red" (how "red"?), or as in "I am a tall person" (how tall is a "tall" person?).

The classic tool for representing uncertainties is Probability Theory, as formulated by Thomas Bayes in the late 18th century. Probabilities translate uncertainty into the lingo of statistics. One can translate "I think that Italy will win the next World Cup" into a probability by examining how often Italy wins the World Cup, or how many competitions its teams have won over the last four years. One can then express a personal feeling in probabilities, as all bookmakers do. Bayes' theorem and other formulas allow one to draw conclusions from a number of probable events.

Technically, a probability simply measures "how often" an event occurs. The probability of getting tails is 50% because if you toss a coin you will get tails half of the times. But that is not the way we normally use probabilities: we use them to express a belief. A proponent of probabilities as a measure of somebody's preferences was the US mathematician Leonard Savage who in the 1950s thought of the probability of an event as not merely the frequency with which that event occurs, but also as a measure of the degree to which someone believes it "will" happen.

The problems with probabilities are computational. Bayes' theorem, the main tool to propagate probabilities from one event to a related event, does not yield intuitive conclusions. For example, the accumulation of evidence tends to lower the probability, not to increase it. Also, the sum of the probabilities of all possible events must be one, and that is also not very intuitive. Our beliefs are not consistent: try assigning probabilities to a complete set of beliefs (e.g., probabilities of winning the World Cup for each of the countries of the world) and see if they add up to 100%. In order to satisfy the postulates of probability theory, one has to change her belief and make them consistent, i.e. tweak them so that the sum is 100%.

Bayes rule ("the probability of a hypothesis being true is proportional to the initial belief in it, multiplied by the conditional probability of an observational data, given that prior probability") would be very useful to build generalizations (or induction), but, unfortunately, it requires one to know the initial belief, or the "prior" probability, which, in the case of induction, is precisely what we are trying to assess.

In 1968 mathematicians Glenn Shafer and Stuart Dempster devised a "Theory of Evidence" aimed at making Probability Theory more plausible.

They introduced a "belief function" which operates on all subsets of events (not just the single events). In the throwing of a dice, the possible events are only six, but the number of all subsets is 64 (all the combination of two sides, three sides, four sides and five sides). In their theory the sum of the probabilities of all subsets is one, the sum of the probabilities of all the single events is less than one. Dempster-Shafer's theory allows one to assign a probability to a group of events, even if the probability of each single event is not known. Indirectly, Dempster-Shafer's theory also allows one to represent "ignorance", as the state in which the belief in an event is not known (while the belief in a set it belongs to is known). In other words, Dempster-Shafer's theory does not require a complete probabilistic model of the domain.

An advantage (and a more plausible behavior) of evidence over probabilities is its ability to narrow the hypothesis set with the accumulation of evidence.

Fuzzy Logic

One of the major breakthroughs in inexact reasoning came in 1965 when the Azerbaijani mathematician Lotfi Zadeh invented "Fuzzy Logic". Zadeh applied Lukasiewicz's multi-valued logic to sets. In a multi-valued logic, propositions are not only true or false but can also be partly true and partly false. A set is made of elements. Elements can belong to more than one set (e.g., I belong both to the set of authors and to the set of Italians) but each element either belongs or does not belong to a given set (I am either Italian or not). Zadeh's sets are "fuzzy" because they violate this rule. An element can belong to a fuzzy set "to some degree", just like Lukasiewicz's propositions can be true to some degree (and not necessarily completely true).

The main idea behind Fuzzy Logic is that things can belong to more than one category, and they can even belong to opposite categories, and that they can belong to a category only partially. For example, I belong both to the category of good writers and to the category of bad writers: I am a good writer to some extent and a bad writer to some other extent. In more precise words, I belong to the category of good writers with a certain degree of membership and to the category of bad writers with another degree of membership. I am not fully into one or the other. I am both, to some extent.

Fuzzy Logic goes beyond Lukasiewicz's multi-valued logic because it allows for an infinite number of truth values: the degree of "membership" can assume any value between zero and one.

Zadeh's theory of fuzzy quantities implicitly assumes that things are not necessarily true or false, but things have degrees of truth. The degree of truth is, indirectly, a measure of the coherence between a proposition about the world and the state of the world. A proposition can be true, false, or… vague with a degree of vagueness.

Fuzzy Logic can explain paradoxes such as the one about removing a grain of sand from a pile of sand (when does the pile of sand stop being a pile of sand?). In Fuzzy Logic each application of the inference rule erodes the truth of the resulting proposition.

Fuzzy Logic is also consistent with the principle of incompatibility stated at the beginning of the 20th century by the French physicist Pierre Duhem: the certainty that a proposition is true decreases with any increase of its precision. The power of a vague assertion rests in its being vague: the moment we try to make it more precise, it loses some of its power. A very precise assertion is almost never certain. For example, "today is a hot day" is certainly true, but its truth rests on the fact that I used the very vague word "hot". If now I restate it as "today the temperature is 36 degrees", the assertion is not certain anymore. Duhem's principle is analogous to Heisenberg's principle of uncertainty: precision and uncertainty are inversely proportional. Fuzzy Logic models vagueness and reflects this principle.

While mostly equivalent to Probability Theory (as proven by the US mathematician Bart Kosko), Fuzzy Logic yields different interpretations. Probability measures the likelihood of something happening (e.g., whether it is going to rain tomorrow). Fuzziness measures the degree to which it is happening (e.g., how heavily it is raining today). And, unlike probabilities, Fuzzy Logic deals with single individuals, not populations. Probability theory tells you what are the chances of finding a tall person in a crowd, whereas fuzzy logic tells you to what degree that person is tall.

Technically, a fuzzy set is a set of elements that belong to a set only to some extent. Each element is characterized by a degree of membership. An object can belong (partially) to more than one set, even if they are mutually exclusive, in direct contrast with one of the pillars of classical Logic: the "law of the excluded middle". Each set can be a subset of another set with a degree of membership. A set can even belong (partially) to one of its parts. Degrees of membership also imply that Fuzzy Logic admits a continuum of truth values from zero to one, unlike classical Logic that admits only true or false (one or zero).

In Kosko's formulation, a fuzzy set is a point in a unitary hypercube (a multi-dimensional cube whose faces are all one). A non-fuzzy set (a traditional set) is one of the vertexes of such a cube. The paradoxes of classical Logic occur in the middle points of the hypercube. In other words, paradoxes such as the liar's or Russell's can be interpreted as "half truths" in the context of Fuzzy Logic.

A fuzzy set's entropy (which could be thought of as its "ambiguity") is defined by the number of violations of the law of non-contradiction compared with the number of violations of the excluded middle. Entropy is zero when both laws hold, is maximum in the center of the hypercube. Alternatively, a fuzzy set's entropy can be defined as a measure of how a set is a subset of itself.

Possibility Theory

Possibility theory (formulated by Zadeh in 1977, and later expanded by French mathematicians Didier Dubois and Henri Prade) developed as a branch of the theory of fuzzy sets in order to deal with the lexical elasticity of ordinary language (i.e., the fuzziness of words such as "small" and "many"), and other forms of uncertainty which are not probabilistic in nature. The subject of possibility theory is the possible (not probable) values of a variable.

Possibility theory is both a theory of imprecision (represented by fuzzy sets) and a theory of uncertainty. The uncertainty of an event is described by a pair of degrees: the degree of possibility of the event and the degree of possibility of the contrary event. The definition can be dually stated in terms of necessity, necessity being the complement to one of possibility.

Its basic axioms are that: 1. the degree of possibility is one for a proposition that is true in any interpretation and is zero for a proposition that is false in any interpretation; 2. the degree of possibility of a disjunction of propositions is the maximum degree of the two. When the degree of necessity of a proposition is one, the proposition is true. When the degree of possibility of a proposition is zero, the proposition is false. When the degree of necessity is zero, or the degree of possibility is one, nothing is known about the truth of the proposition.

Possibility Logic has a graded notion of possibility and necessity.

A Fuzzy Physics?

Unlike Probability theory, Fuzzy Logic represents the real world without any need to assume the existence of randomness. For example, relative frequency is a measure of how a set is a subset of another set.

Many of Physics' laws are not reversible because otherwise causality would be violated (after a transition of state probability turns into certainty and cannot be rebuilt working backwards). If they were expressed as "ambiguity", rather than probability, they would be reversible, as the ambiguity of an event remains the same before and after the event occurred.

Fuzziness is pervasive in nature (everything is a matter of degree), even if science does not admit fuzziness. Even Probability Theory still assumes that properties are crisp, while in nature they rarely are.

Furthermore, Heisenberg's "uncertainty principle" (the more a quantity is accurately determined, the less accurately a conjugate quantity can be determined) can be reduced to the" Cauchy-Schwarz inequality", which is related to "Pythagoras' theorem", which is in turn related to the "subsethood theorem", i.e. to Fuzzy Logic.

One is tempted to rewrite Quantum Mechanics using Fuzzy Theory instead of Probability Theory. After all, Quantum Mechanics, upon which our description of matter is built, uses probabilities mainly for historical reasons: Probability Theory was the only theory of uncertainty available at the time. The result is that Physics has a standard interpretation of the world that is based on population thinking: we cannot talk about a single particle, but only about sets of particles. We cannot know whether a particle will end up here or there, but only how many particles will end up here or there.

The interpretation of quantum phenomena would be slightly different if Quantum Mechanics were based on Fuzzy Logic: probabilities deal with populations, whereas Fuzzy Logic deals with individuals; probabilities entail uncertainty, whereas Fuzzy Logic entails ambiguity. In a fuzzy universe a particle's position would be known at all times, except that such a position would be ambiguous (a particle would be simultaneously "here" to some degree and "there" to some other degree). This might be viewed as more plausible, or at least more in line with our daily experience that in nature things are less clearly defined than they appear in a mathematical representation of them.

The World of Objects

Another aspect of common sense is that it deals with quantities and objects which are a tiny subset of what science deals with (or is capable of dealing with).

The laws of the physical world are relatively simple and few. The daily world of humans is made of a finite set of solid objects that move in space and do not overlap. Each object has a shape, a volume, a mass distribution. For an adequate representation of the physical needs we can get by with Euclides' geometry, an ontology of space-temporal properties and a set of axioms about the way the world works. We need none of the complication of Quantum Mechanics and Relativity Theory. We need no knowledge whatsoever of elementary particles, nuclear and subnuclear forces, and so forth. Life is a lot easier for our senses than it is for laboratory physicists.

What we need to know in order to survive is actually a lot less than what we need to know in order to satisfy our intellectual curiosity. We never really needed to know the gravitational laws in order to survive, as long as we were aware that objects tend to fall to the ground unless we put them on a table or in a pocket or hang them on a wall. We never really needed to be informed of the second law of Thermodynamics, as long as we realized that they can break, but they do not fix themselves.

The US computer scientist Ernest Davis compiled a list of common sense domains. First, we have physical quantities, such as weight or temperature. They have values. And their values satisfy a number of properties: they can be ordered, they can be subdivided in partially ordered intervals, they can be assigned signs based on their derivatives, their relations can be expressed in the form of transition networks, their behavior can be expressed in the form of qualitative differential equations. Then we have time and space. Time operators usually operate in a world of discrete, self-contained situations and events. Space entails concepts of distance, containment, overlapping, boundaries. Physics, in the view of common sense, is a domain defined by "qualitative" rather than quantitative laws, which express the behavior of physical quantities in the context of those temporal and spatial concepts. To this scenario one must add propositional attitudes (specifically the relationship between belief and knowledge), actions (the ability to plan) and socializing (speech acts). Equipped with this basic idea of the world, an agent should be able to go about its environment and perform intelligent actions.

The Measure Space

The British computer scientist Pat Hayes ("Naive Physics Manifesto ", 1978) introduced the "measure space" for a physical quantity (length, weight, date, temperature, etc.). A measure space is simply a space in which an ordering relationship holds. Measurement spaces are usually conceived as discrete spaces, even if the quantities they measure are in theory continuous. In common use, things like birth dates, temperatures, distances, heights and weights are always rounded.

For example, the height of a person is usually measured in whole centimeters (or inches), and omitting the millimeters, and it can be safely assumed that only heights over one meter and less than two meters are possible. This means that the measure space for people's height is the set of natural numbers from 100 (centimeters) to 200 (centimeters). The measure space for driving speed can reasonably be assumed to be the set of numbers from 0 to 160 (kilometers per hour). The measure space for a shirt's size is sometimes limited to four values: small, medium, large, very large. The measure space for jeans' size is a (very limited) set of pairs of natural numbers. The measure space for the age of a person is the set of natural numbers from 1 to 130. The measure space for the date of an historical event is the set of integer numbers from -3,000 (roughly the time when writing was invented) to the number of the year we live in. And so forth.

A measure space is a discrete representation of a continuous space that takes into account only the significant values that determine boundaries of behavior.

Hayes' program was more ambitious than just measurement spaces. Hayes set out to write down in the language of Predicate Logic everything that we take for granted about the world, all of our common-sense knowledge about physical objects. For example, we know that water is contained in something and that, if it overflows, it will run out, but it will not run upward. We know that wood floats in water but iron sinks. We know that a heavy object placed on top of a light object may crush it. We know that an object will not move if placed on a table, but it will fall if pushed beyond the edge. This is what Hayes called "Naïve Physics" and it is the physics that we employ in our daily lives.

Histories of the World

During the 1980s several techniques were proposed for re-founding Physics on a more practical basis.

John McCarthy's "Situation Calculus" ("Situations, actions, and causal laws", 1963) represents temporally limited events as "situations" (snapshots of the world at a given time), by associating a situation of the world (a set of facts that are true) to each moment in time. Actions and events are represented mathematically as mathematical functions from states to states. An interval of time is a sequence of situations, a "chronicle" of the world. The history of the world is a partially ordered sequence of states and actions. A state is expressed by means of logical expressions that relate objects in that state. An action is expressed by a function that relates each state to another state. The property of states is permanence, the property of actions is change. Each situation is expressed by a formula of first-order Predicate Logic. The advantage of this logical apparatus is that causal relations between two situations can be computed.

The elementary unit of measure for common sense is not the point, but the interval. Which interval makes sense depends on the domain: history is satisfied with years (and sometimes centuries), but birth dates require the day.

Points require Physics' differential equations, but intervals can be handled with a logic of time that deals with their ordering relationship.

Qualitative Reasoning

"Qualitative" reasoning is the discipline that aims at describing a physical system through something closer to common sense than Physics' dynamic equations.

In "Qualitative" Physics, a physical system is conceived as made of parts that contribute to the overall behavior through local interactions, and its behavior is represented inside some variation of Hayes' measure space.

Ultimately, qualitative reasoning is a set of methods for representing and reasoning with incomplete knowledge about physical systems. A qualitative description of a system allows for common-sense reasoning that overcomes the limitations of classical Logic. Qualitative descriptions capture the essential aspects of structure, function and behavior, at the expense of others. Since most phenomena that matter to ordinary people depend only on those essential aspects, qualitative descriptions are enough for moving about in the world.

Several approaches are possible, depending on the preferred ontology: Benjamin Kuipers adopts qualitative constraints among state variables; Johan DeKleer focuses on the devices (pipes, valves, springs) connected in a network of constraints; Kenneth Forbus deals with processes by extending the notion of history. Ultimately, a system's behavior is almost always described by constraint propagation.

DeKleer describes a phenomenon in a measure space through "qualitative differential equations", or "confluences". His "envisionment" is the set of all possible future behaviors.

Forbus defines a "quantity space" as a partially ordered set of numbers. Common sense is interested in knowing that quantities "increase" and "decrease" rather than in formulas yielding the quantities' values in time. In other words, the sign of the derivative is more important than the exact value of a quantity.

Kuipers formalizes qualitative analysis as a sequence of formal descriptions. From the structural description the behavioral description (or "envisionment") can be derived, and from this the functional description can be derived. In his quantity space, besides the signs of the derivatives, what matters most are critical or "landmark" values, such as the temperature at which water undergoes a phase transition. Change is handled by discrete state graphs and qualitative differential equations. A qualitative differential equation is a quadruple of variables, quantity spaces (one for each variable), constraints (that apply to the variables) and transitions (rules to define the domain boundaries). Each of these three frameworks prescribes a number of constraint propagation techniques, which can be applied to a discrete model of the physical system.

Physics is a science of laws of nature which are continuous and exact. Things move because they are subject to these laws. Qualitative Physics is a science of laws of common sense that are discrete and approximate. Things move because other things make them move. Qualitative Physics may not be suitable for studying galaxies and electrons, but can work wonders at analyzing a piece of equipment, a machine, and, in general, a physical system made of components. For example, it has been applied to troubleshooting machines: a model of behavior of a system makes it easier to figure out what must be wrong in order for the system to work the way it is working, i.e. which component is not doing its job properly.

Heuretics

The physical world is only one part of the scenario. There is also the "human" world, the huge mass of knowledge that humans tend to share in a natural way: rain is wet, lions are dangerous, most politicians are crooks and carpets get stained.

"Heuristics" is the proper name for most of what we call common sense. Heuristics is the body of knowledge that allows us to find quick and efficient solutions to complex problems without having to resort to mathematical Logic. Heuristics is, for example, the set of "rules of thumb" that most people employ in their daily lives. The intellectual power of our brain is rarely utilized, as in most cases we can find a rule of thumb that will make it unnecessary. We truly reason only when we cannot find any rule of thumb to help us. A human being who did not know any rules of thumb, who did not have any heuristics, would treat each single daily problem as a mathematical theorem to prove and would probably starve to death before understanding where and how to buy food. We tend to employ heuristics even when we solve mathematical problems. And countless games (such as chess) are about our ability to apply heuristics, rather than mere mathematical reasoning.

The US computer scientist Douglas Lenat developed a global ontology of common knowledge and a set of first principles (or reasoning methods) to work with it. Units of knowledge for common sense are units of "reality by consensus": all the things we know and we assume everybody knows; i.e., all that is implicit in our acts of communication. World regularities belong to this tacitly accepted knowledge. And "regularity" may be a key to understand how we construct and why we believe in heuristics.

Lenat's "principle of economy of communications" states the need to minimize the acts of communication and maximize the information that is transmitted.

Another open issue is whether common sense is learned or innate, or: to what extent it is learned and to what extent it is innate. If it is learned, how is it learned?

In the 1940s the Hungarian mathematician Gyorgy Polya studied how mathematicians solve mathematical problems. Far from being the mechanical procedure envisioned by the proponents of the logistic program, he realized that solving a problem required heuristics. Later, he envisioned "Heuretics", a discipline that would aim at understanding the nature, power and behavior of heuristics: where it comes from, how it becomes so convincing, how it changes over time, etc. One of the intriguing properties of heuristics is, for example, the impressive degree to which we rely on it: the moment we realize that a rule of thumb applies, we abandon our line of reasoning. What makes us so confident about the effectiveness of heuristics? Maybe the fact that heuristics is "acquired effectiveness"? The scope of Heuretics is, ultimately, the scientific study of wisdom.

Stupidity

Artificial Intelligence has been researching human intelligence and machine intelligence, possibly abusing the term "intelligence" from the beginning. It can be educational to focus for a few minutes on the opposite quantity, stupidity. Common sense seems to have a perfectly clear understanding of what stupidity is. Most people would agree at once that some statements are stupid.

"Which is the shortest river in the world?" There is no shortest river, because one can always find a shorter stream of water, all the way down to the leak in your bath tub and to a single drop of water in the kitchen sink. While it makes sense to ask which is the longest river in the world, it makes no sense to ask which is the shortest one. As the length gets shorter, the number of rivers increases exponentially.

" Is everybody here?" The question is stupid because if somebody is missing she won't be able to answer the question.

"Does everybody understand English?" exhibits the same type of stupidity.

What do these questions have in common?

Further Reading
Bobrow Daniel: QUALITATIVE REASONING ABOUT PHYSICAL SYSTEMS (MIT Press, 1985)

Bobrow Daniel: ARTIFICIAL INTELLIGENCE IN PERSPECTIVE (MIT Press, 1994)

Brachman Ronald: READINGS IN KNOWLEDGE REPRESENTATION (Morgan Kaufman, 1985)

Davis Ernest: REPRESENTATION OF COMMON-SENSE KNOWLEDGE (Morgan Kaufman, 1990)

Dubois Didier & Prade Henri: POSSIBILITY THEORY (Plenum Press, 1988)

Dubois Didier, Prade Henri & Yager Ronald: READINGS IN FUZZY SETS (Morgan Kaufmann, 1993)

Engelmore, Robert: BLACKBOARD SYSTEMS (Academic Press, 1988)

Forbus Kenneth & DeKleer Johan: BUILDING PROBLEM SOLVERS (MIT Press, 1993)

Gigerenzer Gerd & Todd Peter: SIMPLE HEURISTICS THAT MAKES US SMART (Oxford Univ Press, 1999)

Gupta Anil & Belnap Nuel: THE REVISION THEORY OF TRUTH (MIT Press, 1993)

Heyting Arend: INTUITIONISM (North Holland, 1956)

Hobbs Jerry & Moore Robert: FORMAL THEORIES OF THE COMMONSENSE WORLD (Ablex Publishing, 1985)

Kandell Abraham: FUZZY MATHEMATICAL TECHNIQUES (Addison Wesley, 1986)

Kosko Bart: NEURAL NETWORKS AND FUZZY SYSTEMS (Prentice Hall, 1992)

Kosko Bart: FUZZY THINKING (Hyperion, 1993)

Kuipers Benjamin: QUALITATIVE REASONING (MIT Press, 1994)

Lenat Douglas: BUILDING LARGE KNOWLEDGE-BASED SYSTEMS (Addison-Wesley, 1990)

Lukaszewicz Witold: NON-MONOTONIC REASONING (Ellis Harwood, 1990)

Marek Wiktor & Truszczynski Miroslav: NON-MONOTONIC LOGIC (Springer Verlag, 1991)

Martin-Lof Per: INTUITIONISTIC TYPE THEORY (Bibliopolis, 1984)

Polya George: MATHEMATICS AND PLAUSIBLE REASONING (Princeton Univ Press, 1954)

Savage Leonard: THE FOUNDATIONS OF STATISTICS (John Wiley, 1954)

Shafer Glenn: A MATHEMATICAL THEORY OF EVIDENCE (Princeton Univ Press, 1976)

Sowa John: PRINCIPLES OF SEMANTIC NETWORKS (Morgan Kaufman, 1991)

Turner Raymond: LOGICS FOR ARTIFICIAL INTELLIGENCE (Ellis Horwood, 1985)

Tversky Amos, Kahnemann Daniel & Slovic Paul: JUDGMENT UNDER UNCERTAINTY (Cambridge University Press, 1982)

Weld Daniel & DeKleer Johan: QUALITATIVE REASONING ABOUT PHYSICAL SYSTEMS (Morgan Kaufman, 1990)

Zimmermann Hans: FUZZY SET THEORY (Kluwer Academics, 1985)

Connectionism and Neural Machines

Artificial Neural Networks

An artificial "neural network" is a piece of software or hardware that simulates the neural network of the brain. Several simple units are connected together, with each unit connecting to any number of other units. The "strength" of the connections can fluctuate from zero strength to infinite strength. Initially the connections and their strengths are set randomly. Then the network is either "trained" or forced to train itself. "Training" a network means using some kind of feedback to adjust the strength of the connections. Every time an input is presented, the network is told what the output should be and asked to adjust its connections accordingly.

For example, the input could be a picture of an apple and the required output could be the string of letters A-P-P-L-E. The first time, equipped with random connections, the network produces some random output. The requested output (A-P-P-L-E) is fed back and the network reorganizes its connections to produce such an output. Another image of an apple is presented as the input and the output is forced again to be the string A-P-P-L-E. Every time this happens the connections are modified to produce the same output even if all images of apples are slightly different. The theory predicts that at some point the network will start recognizing images of apples even if they are all slightly different from the ones it saw before.

Formally: a neural net is a nonlinear directed graph in which each element of processing (each node) receives signals from other nodes and emits a signal towards other nodes, and each connection between nodes has a weight that can vary in time.

A number of algorithms have been proposed for adjusting the strengths of the connections based on the expected output. Such an algorithm must eventually "converge" to a unique and proper configuration of the neural network. The network can continue learning forever, but it must be capable of not forgetting what it has already learned. The larger the network (both in terms of units and in terms of connections) the easier it is to reach a point of stability.

Artificial neural networks are typically used to recognize an image, a sound, a written word. But, since everything is ultimately a pattern of information, there is virtually no limit to their applications. For example, they can be used to build expert systems. An expert system built with the technology of knowledge-based systems (a "traditional" expert system) relies on a knowledge base which represents the knowledge acquired over a lifetime by a specialist. An expert system built with neural-network technology would be a neural network which has been initialized with random values and trained with a historical record of "cases". Instead of relying on an expert, one would rely on a long list of previous cases in which a certain decision was made. If the network is fed this list and "trained" to learn that this long list corresponds to a certain action, the network will eventually start recommending that certain action for new cases that somehow match that "pattern".

Imagine a credit scoring application: the bank's experts use some criteria for deciding whether a business is entitled to a loan or not. A knowledge-based system would rely on the experience of one such expert and use that knowledge to examine future applications. A neural network would rely on the historical record of loans and train itself from that record to examine future applications.

The approach is almost completely opposite, even if it should lead to exactly the same behavior.

Parallel Distributed Computing

One can view a connectionist structure as a new form of computing, a different way of finding a solution to a problem (than searching a solution space). Traditionally, we think of problem solving as an activity in which a set of axioms (of things we know for sure) helps us figure out whether something else is true or false. We derive the "theorem" from the premises through a sequence of logical steps. There is one, well-defined stream of information that flows from the premises to the demonstration of the theorem. This is the approach that mathematicians have refined over the centuries.

On the contrary, a connectionist structure such as our brain works in a non-sequential way: many "nodes" of the network can be triggered at the same time by another node. The result of the "computation" is a product of the parallel processing of many streams of information. There are no axioms and no rules of inference. There are just nodes that exchange messages all the time and adjust their connections depending on the frequency of those messages. No logic whatsoever, no reasoning, no "intelligence" is required. Information does not flow: it gets propagated. Computing (if it can still be called "computing") occurs everywhere in the network, and it occurs all the time.

The obvious reason to be intrigued by connectionist (or "neural") computing is that our brain does it, and, if our brain does it, there must be a reason. Another reason is that this form of computing does have advantages over the logical approach. There are many tasks that would be extremely difficult to handle with Logic, but are quite naturally handled by neural computation. For example, what our brain does best: recognizing patterns (whether a face or a sound).

It has been proven that everything that knowledge-based systems do can be done as well with neural networks.

The idea of connectionism, of computing in a network rather than in a formal system, basically revolutionized the very concept of problem solving. After all, very few real-world problems can be solved in the vacuum of pure logic. From weather forecast to finance, most situations involve countless factors that interact with each other at the same time. One can predict the future only if one knows all the possible interactions.

Computational Models of the Brain

In 1943 the US physiologist and mathematician Warren McCulloch, in cooperation with Walter Pitts, wrote a seminal paper ("A Logical Calculus of the Ideas Immanent in Nervous Activity") that laid down the foundations for a computational theory of the brain. McCulloch transformed the neuron into a mathematical entity by assuming that it can only be in one of two possible states (formally equivalent to the zero and the one of computer bits). These "binary" neurons have a fixed threshold below which they never fire. They are connected to other binary neurons through connections (or "synapses") that can be either "inhibitory" or "excitatory": the former bring signals that keep the neuron from firing, the latter bring signals that want the neuron to fire. All binary neurons integrate their input signals at discrete intervals of time, rather than continuously. The model is therefore very elementary: if no inhibitory synapse is active and the sum of all excitatory synapses is greater than the threshold, then the neuron fires; otherwise it doesn't. This represents a rather rough approximation of the brain, but it can do for the purpose of mathematical simulation.

Next, McCulloch and Pitts proved an important theorem: that a network of binary neurons is fully equivalent to a Universal Turing Machine, i.e., that any finite logical proposition can be realized by such a network, i.e., that every computer program can be implemented as a network of binary neurons. Two most unlikely worlds as that of Neurophysiology and of Mathematics had been linked.

It took a few years for the technology to catch up with the theory. Finally, at the end of the 1950s, a few neural machines were constructed. Frank Rosenblatt's "Perceptron" (1957), Oliver Selfridge's "Pandemonium" (1958), Bernard Widrow's and Marcian Hoff's "Adaline" (1960) introduced the basic concepts for building a neural network. For simplicity purposes a neural network can be structured in layers of neurons, the neurons of each layer firing at the same time after the neurons of the previous layer have fired. The input pattern is fed to the input layer, whose neurons trigger neurons in the second layer, and so forth till neurons in the output layer are triggered at last, and a result is produced. Each neuron in a layer can be connected to many neurons in the previous and following layer. In practice, most implementations had only three layers: the input layer, an intermediary layer and the output layer.

After a little while, each layer has "learned" something, but at a different level of abstraction. In general, the layering of neurons plays a specific role. For example, the wider the intermediate layer, the faster but less accurate the process of categorization, and viceversa.

In many cases, learning is directed by feedback. "Supervised learning" is a way to send feedback to the neural network by changing synaptic strengths so as to reflect the error, or the difference between what the output is and what it should have been; whereas in "unsupervised" learning mode the network is able to learn categories by itself, without any external help. Unsupervised learning became feasible after the introduction of algorithms such as Teuvo Kohonen's Self-Organized Maps (1982), Geoffrey Hinton's and Terry Sejnowski's Boltzmann Machine (1983), Stephen Grossberg's Adaptive Resonance Theory (1987) although the idea dates back to British mathematician Albert Uttley's "Informon" ("The informon: A network for adaptive pattern recognition", 1970).

Whether supervised or not, a neural network can be said to have learned a new concept when the weights of the connections converge towards a stable configuration.

Non-sequential Programming

Neural networks are fundamentally different from the sequential, Von Neumann computer. Information is processed in parallel, rather than sequentially. The network can modify itself (i.e., learn), based on its performance. Information is spread across the network, rather than being localized in a particular storage place. The network as a whole can still function even if a piece of the network is not functioning.

The technology of neural networks promised to lead to a type of computer capable of learning, and, in general, of more closely resembling our brain.

The brain is a neural network that exhibits one important property: all the changes that occur in the connections eventually "converge" towards some kind of stable state. For example, the connections may change every time I see a friend's face from a different perspective, but they "converge" towards the stable state in which I always recognize him as him. Some kind of stability is important for memory to exist, and for any type of recognition to be performed. Neural networks must exhibit the same property if they have to be useful for practical purposes and plausible as models of the brain. Several different mathematical models were proposed in the quest for the optimal neural network.

The discipline of neural networks quickly picked up steam. More and more complex machines were built. Until in 1968 the US mathematician Marvin Minsky proved (or thought he proved) some intrinsic limitations of neural networks. All of a sudden, research on neural networks became unpopular and for more than a decade the discipline languished.

Energy-based Models

In 1982 the US physicist John Hopfield ("Neural Networks And Physical Systems With Emergent Collective Computational Abilities") revived the field by proving the second milestone theorem of neural networks. He developed a model inspired by the "spin glass" material, which resembles a one-layer neural network in which: weights are distributed in a symmetrical fashion; the learning rule is "Hebbian" (the rule that the strength of a connection is proportional to how frequently it is used, a rule originally proposed by the Canadian psychologist Donald Hebb); neurons are binary; and each neuron is connected to every other neuron. As they learn, Hopfield's nets develop configurations that are dynamically stable (or "ultrastable"). Their dynamics is dominated by a tendency towards a very high number of locally stable states, or "attractors". Every memory is a local "minimum" for an energy function similar to potential energy. Hopfield's argument, based on Physics, proved that, despite Minsky's critique, neural networks are feasible.

Hopfield's key intuition was to note the similarity with statistical mechanics. Statistical mechanics translates the laws of Thermodynamics into statistical properties of large sets of particles. The fundamental tool of statistical mechanics (and soon of this new generation of neural networks) is the Boltzmann distribution (actually discovered by Josiah-Willard Gibbs in 1901), a method to calculate the probability that a physical system is in a specified state.

Building on Hopfield's ideas, the British computer scientist Geoffrey Hinton and Terrence Sejnowsky ("Massively parallel architectures for A.I.", 1983) developed an algorithm for the "Boltzmann machine" based on Hopfield's simulated annealing. In that machine, Hopfield's learning rule is replaced with the rule of annealing in metallurgy (start off the system at very high "temperature" and then gradually drop the temperature to zero), which several mathematicians were proposing as a general-purpose optimization rule. In this model, therefore, units update their state based on a stochastic decision rule. The Boltzmann machine turned out to be even more stable than Hopfield's, as it will always ends in a global minimum (the lowest energy state).

Probabilistic reasoning had been introduced into Artificial Intelligence by the Israeli computer scientist Judea Pearl with his "Bayesian networks" (1985).

The "back-propagation" algorithm devised in 1986 by the US psychologist David Rumelhart and Geoffrey Hinton ("Learning Representations By Back-Propagating Errors"), a "gradient-descent" algorithm that is considerably faster than the Boltzmann machine, quickly became the most popular learning rule.

The generalized "Delta Rule" was basically an adaptation of the Widrow-Hoff error-correction rule to the case of multi-layered networks, by moving backwards from the output layer to the input layer. This was also the definitive answer to Minsky's critique, as it proved to be able to solve all of the unsolved problems. Hinton and Rumelhart focused on gradient-descent learning procedures. Each connection computes the derivative, with respect to its strength, of a global measure of error in the performance of the network, and then adjusts its strength in the direction that decreases the error. In other words, the network adjusts itself to counter the error it made. Tuning a network to perform a specific task is a matter of stepwise approximation.

The problem with these methods was that they are cumbersome (if not plain impossible) when applied to deeply-layered neural networks, precisely the ones needed to mimic what the brain does.

Deep Learning

In 1986 Paul Smolensky modified the Boltzmann Machine into what became known as the "Restricted Boltzmann machine", which lends itself to easier computation. This network is restricted to one visible layer and one hidden layer, with units in each layer never connected to units in the same layer.

By the end of the 1980s, neural networks had established themselves as a viable computing technology, and a serious alternative to expert systems as a mechanical approximation of the brain. The probabilistic approach to neural network design had won out.

"Learning" is reduced to the classic statistical problem of finding the best model to fit the data. There are two main ways to go about this. A generative model is a full probabilistic model of the problem, a model of how the data are actually generated (for example, a table of frequencies of English word pairs can be used to generate a "likely" sentence). Discriminative algorithms, instead, classify data without providing a model of how the data are actually generated. Discriminative models are inherently supervised. Traditionally, neural networks were discriminative algorithms.

In 1996 the developmental psychologist Jenny Saffran showed that babies use probability theory to learn about the world, and they do learn very quickly a lot of facts. So Bayes had stumbled on to an important fact about the way the brain works, not just a cute mathematical theory.

Hierarchical Belief Networks

This school of thought merged with another one that was coming from a background of statistics and neuroscience. The Swedish statistician Ulf Grenander (who in 1972 had established the Brown University Pattern Theory Group) fostered a conceptual revolution in the way a computer should describe knowledge of the world: not as concepts but as patterns. His "general pattern theory" provided mathematical tools for Identifying the hidden variables of a data set. Grenander's pupil David Mumford studied the visual cortex and came up with a hierarchy of modules in which inference is Bayesian and it is propagated both up and down ("On the computational architecture of the neocortex II", 1992). a feedforward chain of modules in successively higher The assumption was that feedforward/feedback loops in the visual region integrate top-down expectations and bottom-up observations via probabilistic inference. Basically, Mumford applied hierarchical Bayesian inference to model how the brain works.

Hinton's Helmholtz Machine of 1995 was de facto an implementation of those ideas: an unsupervised learning algorithm to discover the hidden structure of a set of data based on Mumford's and Grenander's ideas. The hierarchical Bayesian framework was later refined with Tai Sing Lee of Carnegie Mellon University ("Hierarchical Bayesian inference in the visual cortex", 2003). These studies were also the basis for the widely-publicized "Hierarchical Temporal Memory" model of the startup Numenta, founded in 2005 in Silicon Valley by Jeff Hawkins, Dileep George and Donna Dubinsky. It was another path to get to the same paradigm: hierarchical Bayesian belief networks.

Deep Belief Networks

In 2006 Hinton ("A Fast Learning Algorithm For Deep Belief Nets") made Deep Belief Networks the talk of the town, basically a generative algorithm for Restricted Boltzmann Machines which suddenly relaunched neural networks and led to new, sophisticated applications to unsupervised learning.

Deep Belief Networks are layered hierarchical architectures that stack Restricted Boltzmann Machines one on top of the other, each one feeding its output as input to the one immediately higher, with the two top layers forming an associative memory. The features discovered by one RBM become the training data for the next one.

DBNs are still limited in one respect: they are "static classifiers", i.e. they operate at a fixed dimensionality. However, speech or images don't come in a fixed dimensionality, but in a (wildly) variable one. They require "sequence recognition", i.e. dynamic classifiers, that DBNs cannot provide. One method to expand DBNs to sequential patterns is to combine deep learning with a "shallow learning architecture" like the Hidden Markov Model.

Another thread in "deep learning" originated with Kunihiko Fukushima's convolutional networks ("Neocognitron - A Self-organizing Neural Network Model for a Mechanism of Pattern Recognition Unaffected by Shift in Position", 1980) that led to Yann LeCun 's second generation convolutional neural networks ("Gradient-Based Learning Applied to Document Recognition", 1998). And Yeshua Bengio's stacked auto-encoders further improved the efficiency of deep learning ("Greedy Layer-wise Training of Deep Networks", 2007).

Meanwhile, in 2006 Osamu Hasegawa introduced Self-Organising Incremental Neural Network (SOINN), a self-replicating neural network for unsupervised learning, and in 2011 his team created a SOINN-based robot that learned functions it was not programmed to do.

Psychological Models

Computational models of neural activity soon proliferated. From the "neural equations" devised in 1961 by the Italian physicist Eduardo Caianiello (""An Outline Of Thought Processes And Thinking Machines") to Stephen Grossberg's non-linear quantitative descriptions of brain processes, the number of mathematical theories on how neurons work almost exceeds the possibility of testing them. Now that the mathematics has been improved to the point of safety, the emphasis is moving towards psychological plausibility. At first the only requirement was that a neural network guaranteed to find a solution to every problem, but soon psychologists started requiring that it did so in a fashion similar to the way the human brain does it. Grossberg's models, for example, are aware of Ivan Pavlov's experiments on conditioning.

Besides proving computationally that a neural network can learn, one has to build a plausible model of how the brain as a whole represents the world. In Teuvo Kohonen's "adaptive maps", nearby units respond similarly, thereby explaining how the brain represents the topography of a situation. His unsupervised architecture, inspired by Carl von der Malsburg's studies on self-organization of cells in the cerebral cortex, is capable of self-organizing in regions. Kohonen assumes that the overall synaptic resources of a cell are approximately constant and what changes is the relative "efficacies" of the synapses.

The British computer scientist Igor Aleksander has attempted to build a neural state machine, "Magnus" (1996), that duplicates the most important features of a human being, from consciousness to emotions.

Neurocomputing

Neural networks belong to a more general class of processing systems, parallel distributed processors, and neurocomputing is a special case of Parallel Distributed Processing, or PDP, whereby processing is done in parallel by a number of independent processors and control is distributed over all processes. All the models for neural networks can be derived as special cases of PDP systems, from simple linear models to thermodynamic models. The axiom of this framework is that all knowledge of the system is in the connections between the processors. This approach is better suited than sequential, Von Neumann computing for pattern matching tasks such as visual recognition and language understanding

A concept is represented not by a symbol stored at some memory location, but by an equilibrium state defined over a dynamic network of locally interacting units. Each unit encodes one of the many features relevant to recognizing the concept, and the connections between units are excitatory or inhibitory inasmuch as the corresponding features are mutually supportive or contradictory. A given unit can contribute to the definition of many concepts.

Neurons vs Symbols

Compared with knowledge-based systems, neural networks offer not only different algorithms but also a different view of mental life. Knowledge-based systems rely on Jerry Fodor's model of cognition: knowledge is represented and then computation is performed on that knowledge yielding some kind of action.

The British philosopher Andy Clark, instead, is an advocate of neural networks and highlights the reasons why neural networks provide a more plausible model for cognition than Fodor's "representations". Clark views neural networks (connectionism in general) as a shift of perspective in the way we view the mind, away from a "static" view of mental representations and towards a fluid view of the cognitive activity of the mind, towards the process, not just the structure.

Jerry Fodor's representational theory of mind was meant to provide an explanation of how thoughts become "causes". Fodor assumes that propositional attitudes ("I believe that", "I hope that", "I fear that", "I desire that") are computations on mental representations (eg, a concept such as "my name is Piero"), which, in turn, can be objects of computation because they are symbolic expressions. Each kind of propositional attitude (eg, "belief") expresses a different kind of role and therefore a different kind of computation. Thus, "I believe that my name is Piero" is different from "I hope that my name is Piero" because the computation performed on the mental representation is different. The human brain knows how to represent and compute because it comes equipped with a "language of thought" that works just like the language of mathematical logic.

Clark does not believe such a language exists in the mind and does not believe that Fodor's vision of the mind can account for the "process" of thinking. In Fodor's model, learning is a secondary phenomenon and is largely independent of the environment. Clark, instead, advocates a model in which learning is a fundamental feature of the mind and learning is largely dependent on the environment. That is precisely the difference between knowledge-based models and connectionist models. Moreover, Fodor clearly distinguishes between the computation and the representation, whereas Clark believes that process and representation are one and the same. In a neural network they are.

Neural networks provide a more plausible model for cognition. Clark highlights three key features of connectionism: superposition, context sensitivity and representational change. Superposition is the ability to represent two things with the same structure: the same neural network can be trained (by changing the weights of its connections) to recognize multiple items. Context sensitivity follows from the fact that those weights encode multiple items and therefore the "representation" of something is automatically context-sensitive. Fodor's symbols are always the same regardless of the context in which they are located (the context is expressed via relationships between symbols), whereas neural networks embody the context of what they represent (the context is expressed internally). Representational change is not only the ability to create new representations (Fodor's models can do that by combining symbolic expressions to create new symbolic expressions) but also the acquisition of new representational capacities. The difference is that the former learns by combining pre-existing, internal expressions, whereas a connectionist model learns when trained by an external environment.

Clark also points to general considerations on biological systems. Complex biological systems have evolved subject to the constraints of "gradualistic holism": the evolution of a complex system is possible only insofar as that system is the last or latest link in a chain of structures, such that at each stage the chain involves only a small change (gradualism) and each stage yields a structure that is itself a viable whole (holism). This is precisely the way neural networks grow: at each point in time a neural network is a working network.

To Clark, the process is the key. One cannot break down or troubleshoot how a network does what it does because it depends on the "process" of learning: just looking at the result of learning is not enough to understand how the network performs the task that it has learned to perform. It is like watching a man ride a bike without having watched how he learned to ride the bike: the process of learning is what explains how he is now capable of riding the bike. If we try to analyze his action of riding the bike, we basically try to reduce the task of riding the bike to a set of symbols, which is a contradiction in terms because, again, riding a bike is not obtained by computing symbols, it is obtained by learning how to ride a bike.

Clark points out that, ideally, neural networks should also be able to undergo what developmental psychologist Karmiloff-Smith calls "redescriptions", or complete reorganizations of knowledge that open up new cognitive abilities and lead to a new developmental stage (as happens during child development).

On the other hand, since they are trained by a set of data that comes from the environment, connectionist systems depend on luck: they can only learn if the set of data includes enough statistical information (technically: associative learning is heavily dependent on the statistical distribution of input data). Our brain somehow learns even in a hostile environment that does not provide enough data about this or that concept, but neural networks fail badly to learn anything unless the set of input data is favorable to the desired training (their success depends on the continued availability of a friendly training environment). Clark's suggestion is that the human mind is a neural network that has evolved over thousands of years and therefore has absorbed huge amounts of innate knowledge. In other words, connectionism is not all there is to human cognition: evolution is another big piece of the story, because it predisposes the network.

While Fodor views concepts as the building blocks of thoughts and as represented by fixed structures and as causing action through their relationships, Clark views a concept as a set of skills that a network learns, and views the effect of those skills as the "behavior" of that network. Folk psychology creates the belief that there are such things as concepts when in reality there are only sets of learned skills. To ascribe a concept to a person is to ascribe a set of skills to that person. The set of skills defines the potential behavior of that person or network. Thus "concepts" are basically an illusion created by the language of folk psychology.

The Road from Neurons to Symbols

Computational models of neural networks have greatly helped in understanding how a structure like the brain can perform. Computational models of cognition have improved our understanding of how cognitive faculties work. But neither group has developed a theory of how neural processes lead to symbolic processes, of how electro-chemical reactions lead to reasoning and thought.

A bridge is missing between the physical, electro-chemical, neural processes and the macroscopic mind processes of reasoning, thinking, knowing, etc., in general, the whole world of symbols. A bridge is missing between the neuron and the symbol. Several philosophers have tried to fill the gap.

The "harmony" theory proposed by the US computer scientist Paul Smolensky is an effort in this direction. Smolensky worked out a theory of dynamic systems that perform cognitive tasks at a subsymbolic level. The task of a perceptual system can be viewed as the completion of the partial description of static states of an environment. Knowledge is encoded as constraints among a set of perceptual features. The constraints and features evolve gradually with experience. Schemata are collections of knowledge atoms that become active in order to maximize what he calls "harmony". The cognitive system is, de facto, an engine for activating coherent assemblies of atoms and drawing inferences that are consistent with the knowledge represented by the activated atoms. A harmony function measures the self-consistency of a possible state of the cognitive system. Such harmony function obeys a law that resembles simulated annealing (just like the Boltzmann machine): the best completion is found by lowering the temperature to zero.

The US philosopher Patricia Churchland aims at a unified theory of cognition and neurobiology, of the computational theory of the mind and the computational theory of the brain. According to her program, the symbols of Fodor's mentalese should be somehow related to neurons, and abstract laws for cognitive processes should be reduced to physical laws for neural processes.

Nonetheless, the final connection, the one between the connectionist model of the brain and the symbol-processing model of the mind, is still missing.

Further Reading
Aleksander Igor: IMPOSSIBLE MINDS (Imperial College Press, 1996)

Anderson James & Rosenfeld Edward: NEURO-COMPUTING (MIT Press, 1988)

Anderson James: NEURO-COMPUTING 2 (MIT Press, 1990)

Anderson James: AN INTRODUCTION TO NEURAL NETWORKS (MIT Press, 1995)

Arbib Michael: THE HANDBOOK OF BRAIN THEORY AND NEURAL NETWORKS (MIT Press, 1995)

Bechtel William & Adele Abrahamsen: CONNECTIONISM AND THE MIND (MIT Press, 1991)

Churchland Patricia: NEUROPHILOSOPHY (MIT Press, 1986)

Clark Andy: MICROCOGNITION (MIT Press, 1989)

Clark, Andy: ASSOCIATIVE ENGINES (MIT Press, 1993)

Davis Steven: CONNECTIONISM (Oxford University Press, 1992)

Grenander, Ulf: GENERAL PATTERN THEORY (Oxford Univ Press, 1993)

Grossberg Stephen: NEURAL NETWORKS AND NATURAL INTELLIGENCE (MIT Press, 1988)

Hassoun Mohamad: FUNDAMENTALS OF ARTIFICIAL NEURAL NETWORKS (MIT Press, 1995)

Haykin Simon: NEURAL NETWORKS (Macmillan, 1994)

Hecht-Nielsen Robert: NEUROCOMPUTING (Addison-Wesley, 1989)

Hertz John, Krogh Anders & Palmer Richard: INTRODUCTION TO THE THEORY OF NEURAL COMPUTATION (Addison-Wesley, 1990)

Hinton, Geoffrey & Sejnowski, Terrence: UNSUPERVISED LEARNING : FOUNDATIONS OF NEURAL COMPUTATION (MIT Press, 1999)

Kohonen Teuvo: SELF-ORGANIZING MAPS (Springer Verlag, 1995)

Levine Daniel: INTRODUCTION TO NEURAL AND COGNITIVE MODELING (Lawrence Erlbaum, 1991)

McClelland James & Rumelhart David: PARALLEL DISTRIBUTED PROCESSING vol. 2 (MIT Press, 1986)

Minsky Marvin: PERCEPTRONS; AN INTRODUCTION TO COMPUTATIONAL GEOMETRY (MIT Press, 1969)

Pearl, Judea: PROBABILISTIC REASONING IN INTELLIGENT SYSTEMS (1988)

Rumelhart David & McClelland James: PARALLEL DISTRIBUTED PROCESSING VOL. 1 (MIT Press, 1986)

Smolensky, Paul: INFORMATION PROCESSING IN DYNAMICAL SYSTEMS (MIT Press, 1986)

Uttley, Albert: INFORMATION TRANSMISSION IN THE NERVOUS SYSTEM (1979)

Language: Minds Speak

The Hidden Metaphysics of Language

Language is obviously one of the most sophisticated cognitive skills that humans possess, and one of the most apparent differences between the human species and other animal species. No surprise, then, that language is often considered the main clue to the secrets of the mind. After all, it is through language that our mind expresses itself. It is with language that we can study the mind.

The US linguist Benjamin Lee Whorf extended the view of his teacher, the German-born Edward Sapir, that language is even more than a tool to speak: it is "thought" itself. Or, better, that language and thought influence each other. Language is used to express thought, but, in turn, language shapes thought. In particular, the structure of the language they speak has an influence on the way its speakers understand the environment. Language influences thought because it contains what Sapir had called "a hidden metaphysics", i.e. a view of the world, a culture, a conceptual system (an opinion already expressed by Alexander von Humboldt in 1836). Language contains an implicit classification of experience. Whorf restated Sapir's view in his principle of "linguistic determinism": grammatical and categorial patterns of language embody cultural models. Every language is a culturally-determined system of patterns that creates the categories by which individuals not only communicate but also think.

For example, the joke "I will see you when (if) i come back from my trip" makes no sense in the German language ("when" and "if" are the same word, "wenn").

The US psychologist Katherine Nelson, whose studies focused on the stages of cognitive development in a child, discovered that language is crucial to the formation of the adult mind: language acts as the medium through which the mind becomes part of a culture, through which the shared meanings of society take over the individualistic meanings of the child's mind. Society takes over the individual mind, and it does so through language. From the moment we were born, the ultimate goal of our mind, through our studying, working, making friends, writing books, etc., was to be social.

The US neurologist Karl Lashley ("The problem of serial order in behavior", 1951) showed that a "syntax" similar to the one for language also exists in actions (physical movement) in general, an intuition that would remain unexplored for decades.

Written language is often the one that is studied, but that can be misleading, because language is, first and foremost, spoken, not written.

Written language was invented when the only experience that humans could record and save was visual experience. Not until the end of the 19th century were humans able to record sound (and to this day we don't have a way to record smell, touch and taste). Written language is a visual representation of a sound phenomenon, a trick to make a recording of sound without using sound. Written language separates words with blanks that don't exist in spoken language. Written language translates a continuous sound into a fragmented visual experience; and it does so by splitting sound into atoms (such as words and sentences). Punctuation helps preserve part of the meaning, but the speed and the tone of voice are largely lost.

A Tool to Shape Minds

This view was inspired by the theory of the mediating language advanced in the 1920s by the Russian psychologist Lev Vygotsky: that language provides a semiotic mediation of knowledge and therefore guides the child's cognitive growth. Cognitive faculties are internalized versions of social processes. This implies that cognition develops in different ways in different cultures. Your mind depends on the cultural conditions of the community that raised you.

The individual is the result of a dialectical cooperation between nature and history, between the biological sphere and the social sphere. An individual is a product of culture (nurture) as well as a product of nature. Children develop under the influence of both biology and society.

Vygotsky insisted on the concept of "zone of proximal development": the difference between the unguided (independent) problem solving skills and the guided (coached) problem solving skills.

Language is a way to organize (internally) the world. But language is also a way to transmit mind to less "mentally-able" individuals and across generations: the by-products of this process of "coaching" are the arts and sciences.

The acquisition of language itself is such a process of transmission of mind: teaching a child to speak is a way of coaching the mind of the child.

Humans solve problems by speaking as well as by using their body and tools.

Vygotsky also realized that the process of "learning" from a coach is mostly unconscious (just like the child is not conscious that s/he is learning to speak). He thought it was a general phenomenon: we become conscious of a function only after we have mastered it by practicing it unconsciously.

Human Language And Animal Language

Language is actually quite widespread in nature in its primitive form of communication (all animals communicate and even plants have some rudimentary form of interaction), although it is certainly unique to humans in its human form (but just like, say, chirping is unique to birds in its "birdy" form).

Language is very much a mirror image of the cognitive capabilities of the animal. Is human language really so much more sophisticated than other animals' languages?

Birds and monkeys employ a sophisticated system of sounds to alert themselves of intruders. The loudness and the frequency are proportionate to the distance and probably to the size of the intruder. Human language doesn't have such a sophisticated way of describing an intruder. Is it possible that human language evolved in a different way simply because we became more interested in other things, than in describing the size and distance of an intruder?

If language is about communicating, why is it that there are multiple human languages and not just one? Why is it so difficult to translate from one language to another? And why do we need translators in the first place? Is there any other animal that needs translators when moving from one territory to another?

There are three levels at which human language operates: the "what", the "where", the "why". What are you doing is about the present. Where are you going is about the future. Why are you going there is about the relationship between past and future. These are three different steps of communication. Organisms could communicate simply in the present, by telling each other what they are doing. This is what most machines do all the time when they get connected. Living organisms also move. Bees dance in order to communicate to other bees the location of food ("where?"). Humans are also interested in motives ("why?") all the time. Without a motive a description often sounds incomplete. It is common in rural Southeast Asia to greet people by asking "what are you doing?" The other person will reply "I am rowing the boat". The next question will be "where are you going?" And the last question will be "why are you going there?" With these three simple questions the situation has been fully analyzed, as far as human cognition goes.

This does not mean that there could not be a fourth level of communication that we humans simply do not exhibit because it is beyond our cognitive capabilities.

There are other features that are truly unique to humans: clothes, artifacts, and, first and foremost, fire. Have you ever seen a lion wear the fur of another animal? Light a fire to warm up? Build a utensil to scratch its back? Why do humans do all of these things? Are they a consequence of our cognitive life, or is our cognitive life a consequence of these skills? One wonders if Sapir-Whorf's principle applies only to language or, ultimately, to all behavior.

Language Changes Minds

Language is a form of communication. The linguistic tradition focused on the mental processes of understanding language, thereby taking a "one-brain" view of communication. But communication, by definition, involves (at least) two participants, i.e. two brains. Communication (and language in particular) is a process between two brains. There is a neural process going on in one of the two brains and language is a means for that neural process to affect the neural process occurring in the other brain. Ultimately, "communication" is about one brain trying to replicate some kind of neural pattern into another brain. Language uses sounds (or written symbols) to induce such a mental replication. Those sounds (symbols) are structured in such a way as to interact with the neural process of the other brain and cause it to create a specific neural pattern (that's what we call "understanding"). This is an error-prone process that requires a lot of interaction, due to the fact that each brain is slightly different. But the goal is to eventually transmit a neural pattern from one brain to another. That pattern could be a scene or a story, if we are "narrating" something, or it could be a belief if we are trying to "convince" someone of something, or a concept if we are trying to explain something. It is a pattern that already exists in our brain and we want to recreate it in the brain of our interlocutor.

Needless to say, this implies that brains are capable of changing their neural patterns based on sound/symbols. This is true of all species: bee brains must be capable of changing their neural patterns based on the dances of other bees.

Naturally, once the pattern (a scene, a story, a concept) has been copied in the other brain, it takes on a life of its own because it interacts with the neural pattern that already inhabited that brain.

This complex interplay of brains must provide some significant evolutionary advantage if it appeared and became widespread among all species.

As the US psychologist James-Mark Baldwin noticed, species capable of learning are better at evolving. If language is such an efficient tool for learning that shapes an entire system of thought in a few years, then it is probably useful for survival and evolution.

Ultimately, language creates minds. We not only speak, but also listen. The listening is no less important than the speaking: the speaking expresses our mind, but the listening shapes our mind.

Communication and Ecology

Communication is two beings that engage in changing each other's brain. That is actually the most natural phenomenon if one views life "top-down" and not "bottom-up". When we think bottom-up, we conceive life as many small beings making up societies and larger and larger entities (ecosystems) and eventually making up the Earth. It actually works the other way around: the Earth existed before life as we know it, and the Earth, at any point in time, is also made of living components such as ecosystems, which are made of societies, which are made of individual beings. It is no surprise that all those ecosystems, societies and individuals are capable of communicating: they are merely "parts" of one giant organism, the Earth. Communicating is their natural state. They are "parts" of the same organism.

Communication (and therefore language) is one of the most basic modes of living beings. When a bird sings in the woods, it is most likely telling other birds about the environment. The slightest disturbance will cause the tune to change. The bird singing in the woods is, therefore, reacting to sounds and smells and sights. The sounds the bird is making are "caused" by the environment and are in harmony with the environment. Those "sounds" communicate to other birds information about the environment. Indirectly it is the environment "talking" to the other birds, i.e. to itself.

Language is more than just sound. Language is sound (or vision, when you are reading) with a structure, and therefore packs more information than just sound. Language carries meaning. This was a crucial invention: that you can use sound as a vehicle to carry more information than the sound itself. Again, the tip probably came from Nature itself: Nature speaks to us all the time. The noise of a river or the noise of an avalanche creates concepts in our minds, besides the representation of those sounds. Brain connections are modified at two levels: first to reflect the stimuli of the noise, and then to reflect what we can infer from the noise. Our brain can learn at two levels: there is a noise in that direction, and it is a river (meaning, for example, water to drink). Stimuli modify connections both at the level of perception and at the level of concepts. Language exploits this simple fact.

(The same is true of cinema, but our bodies are not equipped with an organ to make images the way we are equipped with an organ to make sounds, and the invention of writing required a lot less technological knowledge than television or cinema. However, in the future we may end up carrying our portable image-maker so that we can show what happened in images instead of telling it in words).

Sound is not the only way to communicate. Movement can also communicate. Sound is a particular case of movement.

The environment is a symphony of sounds, smells, sights and movement. Language is but one of the instruments in this symphony.

Phonetics

Syntax studies the structure of language, the fact that only some combinations of words are valid. Semantics studies the meaning that those valid combinations create. Pragmatics studies why we speak the way we speak, the purpose of speaking. These three disciplines can ignore the actual speaker: all English speakers use the same syntax; and semantics is about meanings that all humans share; and pragmatics is about attitudes that all members of a linguistic community share. Whether the speaker is Mary or Lucy does not make a difference for the study of how we combine words, how we express meaning and how we use language.

Phonetics, instead, studies the sounds that we make when we speak; and here the challenge is infinitely more complex because a string of words can be pronounced in an infinitely large number of tones at an infinitely large number of speeds, using an infinitely large number of variations of pauses and accelerations. No two speakers pronounce the same sentence the exact same way, and even the same speaker cannot pronounce the same sentence twice using the exact same sounds. To make matters worse, the acoustic signal of someone's voice speaking to us is not broken down into sentences and words: it tends to be a continuous flow of sounds.

If the study of logic can help understand the "digital" processing that takes place in the brain in order to analyze and interpret a sentence (or to package a new sentence), logic can be of little help in figuring out the "analogic" conversion that takes place in the brain in order to transform a string of sounds into a string of words (e.g. the sound that you make when you read aloud the sentence "the cat chases the mouse" into the string of words "the", "cat", "chases", "the" and "mouse").

Hence the field of phonology. While phonetics studied the physical aspects of speech sound, phonology tries to makes sense of it. In the 1920s the Austrian-Russian linguist Nicholas Trubetzkoy introduced the "phoneme" as the elementary unit of speech, and his associate, the Russian linguist Roman Jakobson, founder in 1926 of the "Prague school" of linguistics, showed that the phoneme is defined by a set of distinctive features ("Observations on the phonological classification of consonants", 1939). Each speech sound results from the combination of distinctive features.

Anomalies of Language

The structure of any natural language is so complex that no machine has been able to fully master one yet. It is hard to believe that a child can learn a language at all. Its complexity should make it impossible at the outset.

If we analyze the way language works, we can draw two opposite conclusions: on one hand, the power of language looks overwhelming, on the other, its clumsiness is frustrating.

On one hand, we know that on an average western languages are about 50% redundant: we would not lose any expressive power if we gave up 50% of our dictionary. We can guess the meaning of most sentences from a fragment of them. We also know that professional translators are able to translate a speech with minimal or no knowledge of the topic the speech is about.

On the other hand, we tend to believe that humans have developed amazing capabilities for communicating: language, writing, even television. However, in reality human communication is rather inefficient: two computers can simply exchange in a split second an image or a text, pixel by pixel or character by character, without any loss of information, whereas a human must describe to another human the image in a lengthy way and will certainly miss out on some details. Two computers could even exchange entire dictionaries, in the event they do not speak the same language. They could exchange in a few seconds their entire knowledge. Humans can only communicate part of the information they have, and even that takes a long time and is prone to misunderstandings.

Furthermore, why is it that we can accurately describe a situation, but not our inner life of emotions? Language is so rich when it comes to the external world, but so poor and inefficient when it comes to my inner life.

The Polish linguist Alfred Korzybski noted that the ability to manufacture symbols gives humans a tremendous advantage (the ability to generalize experience and pass them on to other humans, so they do not need to repeat our mistakes or rediscover what we already discovered), but also a disadvantage, that accounts for many of our social and personal problems: there are fewer words (and concepts) than experiences. This means that we use the same word to describe different situations, objects, or feelings. No two apples are the same, but we use the word "apple" for all of them. Worse: we use the word "apple" even for the drawing of an apple, for the dream of an apple and for the string of characters "a-p-p-l-e", which are completely different objects. We tend to equate situations, objects and feelings that are actually different. We tend to define situations more often by "intension" (the "kind" they belong to) than by "extension" (the unique facts of a situation).

The apparent paradox of redundancy has been explained in many ways. English still adds an "s" to the third person singular of a verb ("he eats") even though English mandates the subject ("he eat" would be perfectly understandable) and only for the third person singular and only for the present tense. Italian mandates that all the words are turned plural ("le tre belle ragazze", which means "the three beautiful girls", where each of those words is a plural, even though "three" should be enough and "la tre bella ragazza" would be understandable). These redundant rules can help with the fundamental ambiguity of language and therefore reduce misunderstandings. Redundancy also helps understand mistakes that people make (maybe "he eat" was just a typo and i really meant "we eat", and the missing "s" at the end is a clue). However, it could also be that the redundancy serves non-linguistic purposes. For example, language may have been more "musical" than it is today, and some redundancy may simply be there because it helped say the same thing in more or less musical tones. More importantly, language has always been a way to determine someone's place of birth. It is surprisingly difficult to imitate a dialect, both in accent and in the prevalent words and expressions. Language may have been a powerful tool to recognize members of the same tribe even before it became a powerful tool to philosophize.

Generative Grammar

Many people speak English. So we know that the English language exists and there must be a way to learn it and speak it. However, there is no definition of what the English language is, or of any other natural language. If you want to find out whether a word is English or not, you have to check a dictionary and hope that the author of that dictionary did not miss any word (in fact, almost all of them do miss some words, as new words are created all the time). If you want to find out whether a sentence is English, the individual words are not enough. A foreign word can actually show up in an English sentence. For example, "Mangiare is not an English word" is a perfectly valid English sentence that everybody understands. Even words that are not words in any language can figure in an English sentence: "Xgewut is not a meaningful word" is an English sentence. What makes a sentence English?

At the beginning of the 20th century, the Swiss linguist Ferdinand de Saussure was asking precisely this kind of a question. He distinguished the "parole" (an actual utterance in a language) from the "langue" (the entire body of the language).

Building on those foundations, the US linguist Noam Chomsky started a conceptual revolution. He was reacting to "structural" linguists, who were content with describing and classifying languages, and to behaviorists, who thought that language was learned by conditioning.

At the time, all scientific disciplines were being influenced by a new propensity towards formal thinking that had its roots as much in Computer Science as in David Hilbert's program of "formal systems" in Logic. Chomsky, basically, extended the idea of formal systems to Linguistics: he realized that the logical formalism could be employed to express the grammar of a language; and that the grammar of a language "was" the specification for the entire language. Chomsky's idea was therefore to concentrate on the study of grammar, and specifically syntax, i.e. on the rules that account for all valid sentences of a language.

His assumption was that the number of sentences in a language is potentially infinite, but there is a finite system of rules that defines which sentences can potentially be built and determines their meaning, and that system of rules is what identifies a language and differentiates it from other languages. That system of rules is the grammar of the language.

One of Chomsky's goals was to explain the difference between "performance" (all sentences that an individual will ever use) and "competence" (all sentences that an individual can utter, but will not necessarily utter). We are capable of saying far more than we will ever say in our entire lifetime. And we understand sentences that we have never heard before. You have probably never seen any of the sentences contained in this book but, hopefully, you understand them all. We can tell right away whether a sentence is correct or not, even when we do not understand its meaning. We do not learn a language by memorizing all possible sentences of it. We learn, and subsequently use, an abstraction that allows us to deal with any sentence in that language. That abstraction is the grammar of the language.

Behaviorists thought that language is learned via a process of conditioning: one learns the meaning of a sentence by being exposed to it and to its meaning. But Chomsky pointed out that virtually no sentence is similar to other sentences we've heard before. You have never read a sentence with these exact words before but, hopefully, you understand the meaning of what I just wrote.

Chomsky therefore argued for a "deductive" approach to language: how to derive all possible sentences of a language (whether they have been used or not) from an abstract structure (its "generative" grammar).

Chomsky also argued for the independence of syntax from semantics: the notion of a "well-formed" sentence in the language is distinct from the notion of a "meaningful" sentence. A sentence can make perfect sense from a grammatical point of view, while being absolutely meaningless (such as "the table eats cloudy books").

Phrase Structure

Language is a set of sentences. Each sentence is a finite string of words from a lexicon. And a grammar is the set of rules that determines whether a sentence belongs to that grammar's language. Moreover, the rules of the grammar are capable of generating (by recursive application) all the valid sentences in that language: the language is "recursively numerable".

When analyzing a sentence (or "parsing" it), the sequence of rules applied to the sentence builds up a "parse tree". This type of grammar, so called "phrase-structure grammar", turns out to be equivalent to a Turing machine, and therefore lends itself to direct implementation on the computer.

The phrase-structure approach to language is based on "immediate constituent analysis": a phrase structure is defined by the constituents of the sentence (noun phrase, verb phrase, etc).

Initially, Chomsky thought that a grammar needs to have a tripartite structure: a sequence of rules to generate phrase structure, a sequence of morpho-phonemic rules to convert strings of morphemes into strings of phonemes, and a sequence of transformational rules that transform strings with phrase structure into new strings to which the morpho-phonemic rules can apply.

Whatever the set of rules, the point was that analyzing language was transformed into a mechanical process of generating more and more formal strings, just like when trying to prove a mathematical theorem. The underlying principle was that all the sentences of the language (which are potentially infinite) could be generated by a finite (and relatively small) number of rules, through the recursive application of such rules. And this fit perfectly well with the Logic-based approach of Artificial Intelligence to simulating the mind.

Transformational Grammar

Chomsky thought that two levels of language were needed: an underlying "deep structure", which accounts for the fundamental syntactic relationships among language components, and a "surface structure", which accounts for the sentences that are actually uttered. The latter gets generated by transformations of elements in the deep structure. For example, "I wrote this book" and "This book was written by me" use the same constituents ("I", "to write", "book") and such constituents are in the same relationship: but one is an active form and the other is a passive form. One gets transformed into the other. Their deep structure is the same, even if their surface structures are different. Many different sentences may exhibit the same deep structure.

The phrase structure produces the "deep structure" of a sentence. That needs to be supplemented by a transformational component and a morpho-phonemic component, which together transform the deep structure into the surface structure of the sentence (e.g. active or passive form).

Technically, the deep structure of a sentence is a tree (the "phrase marker"), that contains all the words that will appear in its surface structure. Understanding language, basically, consists in transforming surface structures into deep structures.

In Chomsky's "standard theory" a grammar is made of a syntactic component (phrase structure rules, lexicon and transformational component), a semantic component (that assigns a meaning to the sentence) and a phonologic component (which transforms it into sounds).

In the end, every sentence of the language is represented by a quadruple structure: the D-structure (the one generated by phrase-structure rules), the S-structure (obtained from the D-structure by applying transformational rules), the P-structure (a phonetic structure) and a "logical form". The logical form of a sentence is the semantic component of its representation, usually in the guise of a translation into first-order predicate logic of the "meaning" of the sentence. These four structures define everything there is to know about the sentence: which grammar rules it satisfies, which transformational rules yield its external aspect, which rules yield the sounds actually uttered by the speaker, and finally the meaning of what is said.

Chomsky's computational approach had its flaws. To start with, each Chomsky grammar is equivalent to a Turing machine. Because of Godel's theorem, the processing of a Turing machine may never come to an end. Therefore, a grammar may never find the meaning of a valid sentence, but we have no evidence that our brain may never find the meaning of a valid sentence in our language. Therefore, some conclude that Chomsky's grammars are not what our brain uses. Also, Chomsky had to explain how we can learn the grammar of our own language: if the grammar is computational in nature, as Chomsky thought, then it can be proved mathematically that no amount of correct examples of sentences are enough to learn a language. It is mathematically impossible for a child to have learned the language she speaks!

Universal Grammar

An important assumption lies at the core of Chomsky's theory. Chomsky claimed that we have some innate knowledge of what a grammar is and how it works. Then experience determines which specific language (i.e., grammar) we will learn. When we are taught a language, we do not memorize each sentence word by word: eventually, we learn the grammar of that language, and the grammar enables us to both understand and utter more sentences than we have ever heard. Somehow our brains refuse to learn a language by memorizing all possible sentences: our brains tend to infer a grammar from all those sentences. Chomsky concluded that our brains are pre-wired to deal with grammars, that there exists some kind of universal linguistic knowledge.

The linguistic ability is inherent as much as arms: we do not learn to have arms, we just have them. Experience simply shapes them.

In 1981, Chomsky introduced the concept of a "universal grammar" to defend his thesis that language is innate: we are born with a brain that is pre-wired to learn language; which language we learn depends on what sentences we are exposed to. Our brain is born with a "universal grammar", a set of universal rules that enable it to deal with language (as an abstract skill). Chomsky's point was that languages are impossibly difficult to learn: however, children routinely learn their home language in a few years. Therefore, their brain must be "ready" to acquire language in a way that no computer is. In other words, what the brain has to learn is not the whole concept of "language", but something smaller and simpler. If the brain contains a "universal grammar", then what we have to learn is not the whole concept of "language" but only the specifics of our home language.

Formally stated, Chomsky decomposes a user's knowledge of language into two components: a universal component (the "universal grammar"), which is the knowledge of language possessed by every human, and a set of parameter values and a lexicon, which together constitute the knowledge of a particular language. The ability to understand and utter language is due to the universal grammar that is somehow encoded in the human genome. A specific grammar is learned not in stages, as Jean Piaget thought, but simply by gradually fulfilling a blueprint that is already in the mind.

Children do not learn, as they do not make any effort. Language "happens" to a child. The child is almost unaware of the language acquisition process. Learning to speak is not different from growing, maturing and all the other biological processes that occur in a child. A child is genetically programmed to learn a language, and experience will simply determine which one. The way a child is programmed is such that all children will learn language the same way.

Language acquisition is not only possible: it is virtually inevitable. A child would learn to express herself even if nobody taught her a language.

Chomsky's belief in innate linguistic knowledge is supported by a mathematical theorem proved in 1967 by Mark Gold ("Language Identification In The Limit"): a language cannot be learned from positive examples only. A grammar could never be induced from a set of the sentences it is supposed to generate. But the grammar can correctly be induced (learned) if there is a (finite) set of available grammars to choose from. In that case the problem is to identify the one grammar that is consistent with the positive examples (with the known sentences), and then the set of sentences can be relatively small (the grammar can be learned quickly). .

In 1977 Chomsky, inspired by the US linguist John Ross, also advanced a theory of "government binding" that would reduce differences between languages to a set of constraints, each one limiting the number of possible variants. Grammar develops just like any other organ in the body: an innate program is started at birth but it is conditioned by experience; still, it is constrained in how much it can be influenced by experience. An arm will be an arm regardless of what happens during growth, but frequent exercise will make its muscles stronger. Ditto for grammar. Growth is deterministic to some extent: its outcome can fluctuate, but within limits.

The Universal Grammar of Human Nature

The US linguist Ray Jackendoff thinks that the human brain works thanks to a crucial amount of innate knowledge. As far as language goes, a universal grammar shared by all human brains constrains what language can possibly be, and therefore helps children's brains learn how a language works. That also explains how children learn sign language: children's brains expect the same organization in sign language that they expect in spoken language.

Processing the phonological structure of language has to do with recognizing words and sentences in that sound wave, and viceversa how to organize the sound wave to express words and sentences. The phonological structure is therefore the interface between thought on one hand and the auditory pattern (input) or the vocal pattern (output) on the other. This is not trivial because the acoustic signal of someone's voice speaking to us is not broken down into sentences and words: it tends to be a continuous flow of sounds.

The experience of spoken language is constructed by the hearer's mental grammar: speech per se is only a meaningless sound wave, that only a hearer equipped with the proper decoding device (the universal grammar) can turn into syntactic structures. Children effortlessly understand what is a word within the continuous noise of spoken language: their brain is hardwired to look for words, and words that create sentences.

In fact, Jackendoff argues that the same argument about universal grammars can be applied to all facets of human experience: all experience is constructed by unconscious genetically determined principles that operate in the brain. Language is not the exception, but instead it fits nicely with the other cognitive faculties, operating under a general property of brain functions. Vision too is controlled and enabled by a mental grammar, by a genetic predisposition to recognize objects and situations: just like we can understand a virtually unlimited set of sentences so we can understand a virtually unlimited set of visual situations. The basic principles of visual perception are innate as much as the basic principles of language.

These same conclusions can be applied to thought itself, i.e. to the task of building concepts. Concepts are constructed by using some innate, genetically determined, machinery, a sort of "universal grammar of concepts". The reason that we are capable of organizing the world into concepts is because we (all human brains) are predisposed to organize the world into concepts and to recognize a similar organization in other people's brains.

Human nature is therefore defined by a set of "unconscious patterns" (or specialized modules) that allow us to think a virtually unlimited number of thoughts, to speak a virtually unlimited number of sentences, to recognize a virtually unlimited number of situations, etc. Without some kind of innate module to guide, direct, prune and so forth our mental life, it would be difficult to speak, see and think.

Language is but one aspect of a broader characteristic of the human brain. The brain contains several modules, each specialized in a cognitive function and each driven by a "universal grammar".

Wedding Biology and Linguistics

In the following years, a number of psychologists, linguists and philosophers corroborated the overall picture of Chomsky's vision.

According to the German linguist Eric Lenneberg, language should be studied as an aspect of our biological nature, in the same manner as anatomy. Chomsky's universal grammar is to be viewed as an underlying biological framework for the growth of language. Genetic predisposition, growth and development apply to language faculties just like to any other organ of the body. Behavior in general is an integral part of an organism's constitution.

Another implication of the standard theory (and particularly of its transformational component) is on the structure of the mind. The transformations can be seen as corresponding to mental processes, performed by mental modules (as in Jerry Fodor's computational theory of the mind), each independent of the others and each guided by elementary principles.

The Canadian psychologist Steven Pinker believes that children are "wired" to pay attention to certain patterns and to perform some operations with words. All languages share common features, suggesting that natural selection favored certain syntactic structures. Pinker identified fifteen modules inside the human mind, organs that account for instincts that all humans share.

Our genetic program specifies the existence and growth of the "language organs", and those organs include at least an idea of what a language is. These organs are roughly the same for all humans, just like hands and eyes are roughly the same. This is why two people can understand each other even if they are using sentences that the other has never heard before.

In biological words, the universal grammar is the linguistic genotype. Its principles are invariant for all languages. The values of some parameters can be "selected" by the environment out of all valid values. This pseudo-Darwinian process is similar to what happens with other growth processes. The model used by Gerald Edelman both in his study of the immune system (the viruses select the appropriate antibodies out of those available) and in his study of the brain (experience selects the useful neural connections out of those available at birth) is quite similar.

A disturbing consequence of this theory is that our mental organs determine what we are capable of communicating, just like our arms or legs determine what movements we are capable of. Just like there are movements that our body cannot possibly make, there are concepts that our language can never possibly communicate.

Languages, Not Only Language

A consequence of Chomsky's hypothesis of the universal grammar is that all human languages (or at least their grammars) must be relatively similar, if they all sprung up from the same universal grammar that is genetically transmitted from brain to brain.

There must exist a relatively simple mechanism by which the brain constructs a specific grammar (say, English) out of the universal grammar. The US linguist Mark Baker tried to identify the similarities among languages and what "parameters" determine whether you speak, say, English or French. Imagine a car that you could personalize so much that it could eventually look completely different from the car of your neighbor, while still being manufactured at the same plant. The "personalization" would consist in selecting "options" such as paint color, body shape, number of doors, etc. At each step of the assembly line, machines would obey one of the options and add a different touch to your car. Something similar occurs in the brain with language, according to Baker.

Baker imagines a "tree" (a hierarchy) of such linguistic parameters. The parameters are arranged according to their power to affect one another. At each junction in the hierarchy, a parameter (or more) determines a way to structure sentences. Below that junction, that parameter is fixed and other parameters are taken into account. Baker's hierarchy of linguistic parameters looks like the periodic table of elements or Carl Linnaeus' classification of animals and plants, but it is different in that it specifies how to "generate" such a classification.

Alas, Baker does not account for the evolution of languages. We do know that Italian evolved from Latin. Italian has wildly different grammatical rules from Latin. How did it happen that children born with a pre-wired brain and later brainwashed to acquire the parameters of the Latin language ended up changing those parameters? In fact, the one thing that seems to change a lot is the grammar itself. Words are relatively similar among languages of the same geographic area, but the grammar can be quite different (as any Italian who studied Latin in high school painfully remembers). Those changes occurred gradually over many generations.

It is possible that all languages descend from the same original language; and that children learn a language in the amount of time that it takes to learn a language, not any faster. These may be clues to another scenario for the origin of language.

No question the brain is born with a structure that facilitates survival on this planet. It is even too easy to claim that so many skills must be "innate". We would not be alive if that were not the case. However, claiming that our brains are "pre-wired" to discover General Relativity would be an exaggeration. It took Albert Einstein to discover General Relativity. Decades from now every human on the planet will master General Relativity. That does not mean that General Relativity was inevitable for all human brains. There may have been an Einstein of communicating, who invented language. And we all might be simply re-learning, generation after generation, that marvelous invention.

The Language Instinct

Pinker spoke of a "language instinct" rather than of a "universal language". He noticed how inherently, and naturally, complex language is. It is virtually impossible for a community of human beings to develop a "simple" language. As Pinker puts it: "there are stone-age societies, but there is no such thing as a stone-age language". All languages, even in the most primitive societies, are complex, convoluted, redundant, full of exceptions, synonyms and ambiguities. His explanation is that humans are equipped with a "language instinct" that makes them reinvent language generation after generation.

The way children learn a language seems to be largely independent of what their parents teach them. For example, children use grammatical forms that they never heard from their parents. The ability of children to learn a language depends on the extent to which their brains are biased towards some kinds of phrases (e.g., noun phrases) and equipped with some kinds of procedures.

Pinker thinks that language is merely a (very limited) medium to pack our thoughts and broadcast them to our fellow humans. In the process, we lose most of the reasoning that went on in our brains. Pinker thinks that the mental representation of what we are saying is way more complex and subtle, and just cannot be expressed in words. Words are an effective way to deliver the essential part of the meaning in a reasonable amount of time. Our thoughts are not verbal: they are mental representations of the kind that resembles Jerry Fodor's "mentalese". And this mentalese is a genetic fact: we inherit it when we inherit human genes. It is universal. We all reason the same way. We think in mentalese, not in English or Chinese. It is only when we have to pack information for another human being that we use the language of our community (e.g., English or Chinese), and in doing so we have to limit our message to what can be said in that language.

Pinker therefore sides with the school of the "physical symbolic processor": a mind is a purely syntactic processor of symbols, and the "intelligence" of that mind arises from processing the symbols, just like the Turing machine is capable of solving (almost) every problem by moving symbols around, without actually "knowing" what problem it is solving (or even that it is solving a problem).

Pinker's point is that mental representation is a powerful invention. Once we equip something with the ability to represent the world through symbols that can be processed in some logical way, that something is suddenly endowed with the power to solve even the most complex of problems. Mental representation sounds indeed miraculous. The symbols are not intelligent. The algorithm to process them is not intelligent. But the result is intelligent. In fact, the result "is" intelligence itself.

Language and thought have little in common, according to Pinker: "knowing a language is knowing how to translate mentalese into strings of words, and viceversa". People without a language are still thinking in mentalese.

Pinker notes that mentalese might actually be simpler than the languages we speak, because it doesn't have to deal with the oddities of spoken language (such as pronouns and indexicals) or with pronunciation. Presumably, Pinker thinks that the complexity of spoken languages evolved because it helps communicate the essential very quickly.

Phonetic perception itself is part of the "language instinct". Sentences are made of words, words of morphemes, and morphemes of phonemes. And phonemes are fundamentally different from the other components of language, because they do not combine according to a grammar: they represent the analogic to digital interface.

At some point in evolution, the mouth and the ear developed additional functions: to help utter sounds and to help to "understand" sounds. Pinker claims that we can hear "words" where there are just sounds because phonetic perception is a sixth sense, another piece in the puzzle of the language instinct. Our brains are hardwired to recognize "meaningful" words out of a stream of "meaningless" sounds (there are actually no meaningless sounds, and, according to Pinker's own theory, words are not really meaningful, but Pinker uses "meaningful" as in "useful for the purpose of reacting to a sentence"). Pinker shows how we use an assembly of organs to create the sounds of sentences. Recognizing a phoneme is much more difficult than dealing with grammar, so much so that no machine has been built yet that can recognize speech the way the human brain does. First of all, there is hardly any separation between words when we speak: there is a continuous flow of sounds. Secondly, different speakers pronounce the same words in different ways. Thirdly, the same speaker can pronounce the same word in different ways (depending on whether she is sleepy or not, angry or not, in a hurry or not). Machines that try to recognize speech have to be "trained" to the voice of a particular speaker, and can generally recognize only a small subset of the vocabulary (usually, only a dozen of words, instead of the tens of thousands that the human brain recognizes effortlessly). Clearly, the wiring of the human brain is the secret to recognizing speech.

There are other features of the human brain that are difficult to replicate with a machine, and they all have to do with "analogic" versus "digital" reasoning. Uttering or listening to speech is a function very similar to grasping an object, another difficult act for a machine. The brain does not simply calculate distance and angle to program the movement of the arm. The movement of the arm and of many other organs, from the joints of the fingers to the muscles of the eyes are continuously refined as the movement itself takes place. A complex process of feedback makes sure that the movement achieves its goal. Pinker shows that something similar happens when we speak: a number of organs cooperate in making the sounds of words, and the sound wave is refined as it is being uttered; and viceversa when the sound wave is being heard. Thus speech belongs to the general class of "motor control". Just like with other kinds of motor control, the "expectation" contributes to the success of the operation. If one "expects" to hear something, then she is more likely to hear it. Since we expect a speaker to make sounds that are words in our language, we are more likely to detect them, even though the words are not coded in the exact same sound wave each and every time. Speech recognition is greatly facilitated if the hearer continuously guesses what the speaker is trying to say. The secret of speech is not in its "digital" workings (how the brain dissects and processes its constituents) but in its "analogic" workings (how the brain continuously readjusts the process to sculpt an output that matches the input).

If all humans are equipped with the same universal grammar, a legitimate question is why are there so many languages instead of just one. Pinker's answer is similar to the answer to the question of why are there so many species of animals if all animals are equipped with the same genetic code: it's the way evolution works, namely variation is an inherent element of evolution. Linguistic variation boosts cultural evolution the same way that genetic variation boosts biological evolution. New languages are born the same way that new species are born, through a process of variation, heredity and isolation. (This similarity had originally been pointed out by Darwin himself). Pinker does not elaborate on the linguistic equivalent of "natural selection", i.e. the role played by the "environment" (which, in the case of language, is the society of other speakers), but language too is subject to environmental pressure. If a child utters a meaningless sentence that brings no benefit (or is even harmful in achieving the goal), that sentence will die out. On the other hand, novel sentences or grammatical constructs or idiomatic expressions that turn out to be very effective are inherited by other speakers and spread throughout the population of speakers. This is the equivalent of what natural selection does to organs of bodies.

Pinker calls it "language instinct". Like all instincts, it must be implemented somewhere in the brain, and that implementation must be dictated by some genes of the human genome (and neurologists proved that language is implemented in the left hemisphere). That is why Pinker thinks that animals cannot speak language, "real" language: they lack the genes, and therefore they lack the brains. They can certainly be trained to recognize and react to certain sounds (although never with the same dexterity of a child) but they lack the "discrete combinatorial system" that would enable them to understand "other" sentences besides the ones they have been trained to react to. Children do not simply repeat the sentences that they have been taught: children come up with their own sentences. What children have learned is "language", not just a few words or a few sentences. That is what animals cannot learn. Animals can learn to react to the sentence "Kiss me" and "Dog", but they cannot understand the sentence "Kiss the dog". Even less likely is that they can understand the sentence "Kiss her". And even less likely to reply to a simple question such as "Why?" Children, instead, rapidly learn to deal with sentences such as "Why aren't you nice to her?" even if they never heard those words before in that specific sequence, as long as they have learned what "nice" means, and what pronouns stand for, and what is expected by a "why".

Neurally speaking, Pinker thinks that human language is ultimately controlled by the neocortex, whereas animal "language" is controlled by the evolutionary older structures in the brain stem and in the limbic system. Humans too have this primitive form of language controlled by the same ancient brain structures, but those are the kind of sounds that cannot be "combinatorially combined", for example a scream of terror or a burst of laughter. Human language is not a combination of these primitive sounds, but a different process altogether of syntax, morphology and phonology that takes place in a different region altogether of the brain.

Case-based Grammars

Either because they did not agree with his vision of the human mind, or because they considered unnatural the limits of his grammars, or because they devised mathematically more efficient models, other thinkers rejected or modified Chomsky's theory.

One powerful idea that influenced many thinkers is that the deep structure of language is closer to the essence of concepts than to the syntax of a specific language. For example, "case-frame grammar" (developed in the late 1960s by the US linguist Charles Fillmore) assumes that each sentence represents explicitly the relationships between concepts and action. Fillmore shifted the emphasis towards "cases". Traditional cases (such as the ones used by German or Latin) are purely artificial, because the same sentence can be rephrased altering the cases of its constituents: "Piero" in "Piero wrote this book" and "this book was written by Piero" appears in two different cases. But its role is always the same, regardless of how we write the sentence. That role is the real "case", in Fillmore's lingo. These cases are universal. They are not language-specific. My relationship towards the book i wrote is the same in every language of the world, regardless of how a specific syntax allows me to express such relationship. Fillmore concluded that a universal underlying set of case-like relations plays a fundamental role in determining syntactic and semantic relations in all languages.

In Fillmore's grammar, therefore, a sentence is represented by identifying such cases. Sentences that deliver the same meaning with different words, but that describe essentially the same scene, get represented in the same way, because they exhibit the same items in the same cases.

Fillmore's approach started a whole new school of thought.

Drawing from the Aristotelian classification of state, activity and eventuality, the US linguist David Dowty proposed that the modal operators "do", "become" and "cause" be used as the foundations for building the meaning of every other verb. Within a sentence, the various words have "roles" relative to the verb. A thematic role is a set of properties that are common to all roles that belong to that thematic role. A thematic role can then be seen as the relationship that ties a term with an event or a state. And this allows one to build a mathematical calculus (a variant of the lambda calculus) on thematic roles.

Likewise, Ray Jackendoff proposed that the meanings of all verbs be reduced to a few space-time primitives, such as "motion" and "location".

Conceptual Dependency

The culmination of this school was, in the 1970s, Roger Schank's "conceptual dependency" theory, whose tenet is that two sentences whose meaning is equivalent must have the same representation. This goal can be achieved by decomposing verbs into elementary concepts (or semantic primitives).

Sentences describe things that happen, i.e. actions. While every language provides for many different actions (usually expressed as verbs), it turns out that most of them can be defined in terms of simpler ones. For example, "to deliver" is a combination of "to move" and other simpler actions. In other words, a number of primitive actions can be used to form all complex actions. In analyzing language, one can therefore focus on those primitive actions. Or, equivalently, a verb can be decomposed in terms of more primitive concepts.

Each action entails roles which are common to all languages. For example, "to move" requires an agent who causes the move to happen, an object to be moved, an old location and a new location, possibly a timeframe, etc. Sometimes the roles are not explicitly stated in a sentence, but they can be derived from the context. Whether they are stated in the sentence or implicit in the context, those roles always exist. And they exist for every language that has that concept. "To move" may be translated with different words in different languages, but it always requires a mover, an object, etc. An important corollary is that any two sentences that share the same meaning will have exactly the same representation in conceptual dependency, regardless of how much is left implicit by each one, regardless of how each is structured. If they refer to the same mover and the same object and the same location and the same timeframe, two different sentences on "moving" will have identical representations.

Conceptual dependency reveals things that are not explicit in the surface form of the utterance: additional roles and additional relations. They are filled in through one's knowledge of lexical semantics and domain heuristics. These two components help infer what is true in the domain.

Conceptual dependency represented a major departure from "Chomskyan" analysis, which always remained relatively faithful to the way a sentence is structured. Schank's analysis considers negligible the way words have been assembled and shifts the emphasis on what is being described. If nothing else, this is presumably closer to the way our memory remembers them.

Semantics

In Chomsky's linguistic world, the "meaning" of a sentence was its logical form. At the end of the process of parsing a sentence, a logical translation would be produced which allowed for mathematical processing, and that logical form was considered to be the meaning of the sentence.

Unfortunately, syntax is ambiguous.

Sentences like "Prostitutes appeal to Pope" or "Soviet virgin lands short of goal again" (actual newspaper headlines reported by Keith Devlin) are "ambiguous". Language does that. In every language one can build a sentence that is perfectly valid but not clear at all. Solving ambiguities is often very easy. If the second sentence is encountered in the context of the development of Siberia, one may not even notice the ambiguity. The context usually solves the ambiguity.

A related problem is that of "anaphora". The sentence "He went to bed" is ambiguous in a different way but still ambiguous: technically speaking, "he" could be any of the 3 billion males who live on this planet. In practice, all we have to do is read the previous sentences to find out who "he" is. The context, again, helps us figure out the meaning of a sentence.

Not to mention expressions such as "Today is an important day" or "Here it is cold": when and where are these sentences occurring?

Because of linguistic phenomena like ambiguity and anaphora, understanding a discourse requires more than just figuring out the syntactic constituents of each sentence. In fact, even understanding the syntactic constituents may require more than syntax: the "lies" in "Reagan wins on budget but more lies ahead" is a noun (plural of "lie") or a verb (third person of "to lie")?

The scope of semantics lies beyond the single word and the way the words relate to each other.

In the 1960s the US linguist Jerrold Katz provided one of the most extensive studies on semantics. His basic tenet is that two components are necessary for a theory of semantics. The first one is a dictionary, which provides for every lexical item (i.e., for every word) a phonological description, a syntactic classification ("grammatical marker", e.g. noun or verb) and a specification of its possible distinct senses ("semantic marker", e.g. "light" as in color and "light" as the opposite of heavy). The second one is a set of " projection rules" that determine how the meaning of a sentence can be composed from the meaning of its constituents. Projection rules, therefore, produce all valid interpretations of a sentence.

The Math of Grammars

First-order predicate logic, the most commonly used logic, may be too limited to handle the subtleties of language.

The "generalized phrase-structure grammar" pioneered by the British linguist Gerald Gazdar, for example, makes use of "Intensional Logic", which is a variant of the "lambda calculus". Gazdar abandoned the transformational component and the deep structure of Chomsky's model and focused on rules that analyze syntactic trees rather than generate them. The rules translate natural language sentences in an intensional-logic format. This way the semantic interpretation of a sentence can be derived directly from its syntactic representation. Gazdar defined 43 rules of grammar, each one providing a phrase-structure rule and a semantic-translation rule that show how to build an intensional-logic expression from the intensional-logic expressions of the constituents of the phrase-structure rule. Gazdar's system was fundamentally a revision of Katz's system from predicate logic to intensional logic.

The US mathematician Richard Montague developed the most sophisticated of intensional-logic approaches to language. His intensional-logic system employed all sorts of logical tools: type hierarchy, higher-order quantification, lambda abstraction for all types, tenses and modal operators; and its model theory was based on coordinate semantics.

In this version of intensional logic the sense of an expression determines its reference. The intensional-logic formula makes explicit the mechanism by which this can happen.

Reality consists of two truth values, a set of entities, a set of possible worlds and a set of points in time. A function space is constructed inductively from these elementary objects.

Montague's logic determines the possible sorts of functions from possible "indices" (sets of worlds, times, speakers, etc.) to their "denotations" (or extensions). These functions represent the sense of the expression. In other words, sentences denote extensions in the real world. A name denotes the infinite set of properties of its reference. Common nouns, adjectives and intransitive verbs denote sets of individual concepts, and their intensions are the properties necessarily shared by all those individuals.

Through a rigorously mechanical process, a sentence of natural language can be translated into an expression of intensional logic. The model-theoretic interpretation of this expression serves as the interpretation of the sentence.

Rather than proving a semantic interpretation directly on syntactic structures, Montague provides the semantic interpretation of a sentence by showing how to translate it into formulas of Intensional Logic and how to interpret semantically all formulas of that logic.

Montague assigns a set of basic expressions to each category and then defines 17 syntactic rules to combine them to form complex phrases. The translation from natural language to intensional logic is then performed by employing a set of 17 translation rules that correspond to the syntactic rules. Syntactic structure determines semantic interpretation.

Montague's work was based on the idea of "categorial grammars" pioneered by the German mathematician Yehoshua Bar-Hillel ("A Quasi-arithmetical Notation for Syntactic Description", 1953), the man who had organised the first conference on machine translation in 1952. Categorial grammar is built up from two primitive categories: noun phrase and verb phrase. A sentence is composed of a noun phrase (Piero, the red apple, the president of the US) and a verb phrase (wrote this book, is rotting, has canceled his trip). They can be related in an arithmetic way by using the same rules of fractions: a verb phrase VP is the sentence S divided by the noun phrase NP. Categorial grammars provide a unity of syntactic and semantic analyses.

Montague's semantics is truth-conditional (to know the meaning of a sentence is to know what the world must be for the sentence to be true, or the meaning of a sentence is the set of its truth conditions), model-theoretic and uses possible worlds (the meaning of a sentence depends not just on the world as it is but on the world as it might be, i.e. on other possible worlds).

Natural Semantics

The Polish linguist Anna Wierzbicka was the originator of "natural semantics".

She regarded language as a tool to communicate meaning, and semantics as the study of meaning encoded in language. To her, syntax is a piece of semantics.

Corresponding to the three types of tools employed by language to convey meaning (words, grammatical constructions and "illocutionary" devices), linguistics can be divided into lexical semantics, grammatical semantics and illocutionary semantics. To her the division into syntax, semantics and pragmatics makes no sense because every element and aspect of language carries meaning. Meaning is an individual's interpretation of the world. It is subjective and depends on the social and cultural context. Therefore, semantics encompasses lexicon, grammar and illocutionary structure.

Wierzbicka's project was to build semantics from elementary concepts. There exist a broad variety of semantic differences among languages (even emotions seem to be cultural artifacts), but she identified a few semantic primitives shared by all languages. Such universal semantic primitives make up a semantic meta-language that could be used to explicate all other concepts in all languages.

Cognitive Grammar

The US linguist Ronald Langacker, one of the originators of cognitive linguistics (whose first conference was organized in 1989), reacted against the prevailing view that language is a self-contained system that can be studied in isolation. He opposed the view that grammar was distinct from lexicon and semantics, and that the meaning of a sentence could be expressed in mathematical logic. Langacker believed that language cannot be separated from cognition, that semantics is about concepts, and that semantic analysis is conceptual analysis. Ultimately, he believed that language is psychology and neurology as much as it is linguistics.

Noting that grammar is simply a way to refer symbolically to concepts, i.e. that grammar is a symbolic element connecting phonology (the sounds of speech) and concepts, Langacker recast grammar as an extension of the lexicon. Grammar is an "inventory of symbolic resources". Grammatical units have a meaning, just like the items of a lexicon have meaning. This "meaning" cannot be merely a truth condition or a combination thereof, because that meaning is related to the whole cognitive process of understanding/speaking language, i.e. to a cognitive domain.

For example, the class of nouns refers to a kind of cognitive processing, and that is its meaning, whereas the class of verbs refers to a different kind of cognitive processing, and that is its meaning. And different classes of nouns (e.g., count nouns as opposed to mass nouns) refer to different kinds of "noun" cognition.

Any item in the lexicon (any word) refers to a kind of cognitive processing, which is its meaning.

Langacker admits only three kinds of units: semantic (the concepts), symbolic (grammar, lexicon, morphology) and phonological (the sounds). The symbolic units connect units of the other two kinds.

At the same time, the form used to construct the concept is also "meaningful". One can create a content using many different forms of language. Langacker used the term "imagery" to refer to how content is structured. By definition, a grammar already forces constraints on the "images" that content can assume. Each grammar already limits the universe of imagery that is available to the language user.

Rather than sentences and grammatical rules, Langacker's grammar is built on image schemas, which are schemas of visual scenes.

Again, only a semantic and a phonological component are necessary, mediated by a symbolic component. This approach directly reflects the semiological function of language: to build symbols for concepts (semantics) by means of sounds (phonology). Grammar reduces to these symbolic relationships between semantic structures and phonological structures.

A speaker's linguistic knowledge is contained in a set of cognitive units, which are originated by a process of reinforcement of recurring features (or "schematization"), or, identically, by a process of patterns of neural activity. These units are, therefore, grounded in daily experience and are employed by speakers in automatic fashion: a unit is a whole that does not need to be broken down into constituents to be used. Phonological units, for example, range from the basic sounds of language (such as the "t" of the English language or the "r" of the French language) to familiar phrases and proverbs.

Units form a hierarchy, a schema being instantiated in sub-schemas. A linguistic category may be represented by a network of quite dissimilar schemas, clustered around a prototype. A grammar is but an inventory of such units.

Nouns and verbs are central to grammatical structure because of the archetypal status of a cognitive model whose elements are space, time, matter and energy. That is a world in which discrete physical objects move around in space thanks to some form of energy, in particular the one acquired through interactions with other objects. Matter spreads over space and energetic interactions occur over time. Objects and interactions are instantiated, respectively, in space and time. Objects and interactions are held to be the prototypes, respectively, for the noun and verb grammar categories. These categories differ primarily in the way they construe a situation, i.e. their primary semantic value is "imagic", has to do with the capability to construe situations.

Langacker took issue with the "Chomskyan" view that language is an infinite set of well-formed sentences or any other algorithm-generated set. To him, a language is a psychological phenomenon that eventually resides in neural activity. Chomsky's generative grammar is merely a platonic ideal.

Mental Spaces

A similar change in perspective was advocated by the French linguist Gilles Fauconnier.

Fauconnier's focus was on the interaction between grammar and cognition, i.e. into the interaction between syntax/semantics and "mental spaces". The mind is capable of making connections between domains and Fauconnier investigates the kinds of cognitive connections that are possible: pragmatic functions (such as that between an author and her book), metonymy, metaphor, analogy, etc. Some domains are cognitively accessible from others and meaning is to be found in these interactions.

A basic tenet of Fauconnier's theory is that linguistic structure reflects not the structure of the world but the structure of our cognitive life.

The idea is that, as the speaker utters one sentence after the other, she is in fact constructing mental spaces and the links among them, resulting in a network of mental spaces. Language builds the same kind of mental spaces from the most basic level of meaning construction all the way up to discourse and reasoning. While logic-based semantics (whether Chomsky's or Montague's) assumed that language provides a meaning that can be used for reasoning, Fauconnier maintained that mental spaces facilitate reasoning.

Furthermore, mental spaces allow for alternative views of the world. Fauconnier thinks that the mind needs to create multiple cognitive spaces in order to engage in creative thought.

Cognitive Linguistics

The US linguist George Lakoff is critical of Chomsky's theory on philosophical grounds: Chomsky's theory belongs to the old logical-analytical tradition, because Chomsky embraced logical formalism and several of Descartes' assumptions while neglecting how thinking and language rest on bodily experience. Lakoff does not believe in an innate, universal grammar. Lakoff does not believe that the structure of language is independent of meaning.

Lakoff's "cognitive linguistics" rests on the opposite assumption that language (like anything else in mental life) is grounded in our bodily experience. Language is embodied, which means that its structure reflects our bodily experience. Syntax is a consequence (not a prerequisite) of concepts. Our bodily experience creates concepts that are then abstracted into syntactic categories. Syntax is a direct consequence of our bodily experience, not an innate property. It is shared (to some degree) by all humans for the simple reason that we all share roughly the same bodily experience.

Chomsky Revisited

In 1986 Noam Chomsky, aware of the shortcomings of his generative theory of language, introduced a new theory of language, the theory of "Principles and Parameters", later (1995) renamed "Minimalism".

Chomsky recognized that there might be no universal grammar, just a circuit in the brain that is more or less plastic: change the connections and you get one or the other language. Instead of innate knowledge of language, Chomsky proposed that the brain comes equipped with a virtually infinite set of concepts. There are no "rules" of grammar as such, but there are associations between sounds and concepts: we learn a concept when we make the connection with a sound. Basically, we "rediscover" concepts that we have always unconsciously known (they have always been in our mind, since, presumably, prehistoric times).

Chomsky basically claimed that the "rules" of grammar are only a consequence, a side-effect, of the way language works. One could come up with a set of rules of how a muscle or a stomach works, but it is not that the brain has rules on how to run the muscle or the stomach: the rules are a way to explain what actually happens. The key is to discover the mechanism that generates those apparent "rules" of behavior (and their countless exceptions).

Most linguists simply neglect history and the fact that we are a species capable of learning and of transmitting knowledge. Were we a species that does not change over the centuries, Chomsky's original theory of language might have worked just fine. Alas, we keep changing our culture and our behavior, and we instruct our children to maintain our changes. Whatever human phenomenon we observe we are bound to be confused by our own messing with it over the millennia. There might indeed be simple mechanisms that explain language, but those mechanisms are probably perturbed by the fact that humans continuously change their own culture, including their own language. Thus, at every point in time, one can find countless exceptions to every rule. Those exceptions are probably a sign that language is in progress, changing as we observe it. Imagine if you had to study the behavior of a machine while the machine is being dismantled and rebuilt. That is what we do when we study any human phenomenon.

A study of the history of language might show that there are many more regularities than one supposes. Irregular verbs probably have a reason to be what they are (they may have been regular in the past, according to a long-forgotten rule). Words may be derived from very simple sounds. Idiomatic expressions may be based on bodily features. And so forth. If one studies the history of a language, there might be simple explanations for every "odd" feature of it.

Language as a By-product

The US psycholinguist Roger Brown refined the view of Jean Piaget's "constructivism" that language acquisition follows the acquisition of cognitive skills. According to Brown, language is acquired via a "law of cumulative complexity". Language follows the acquisition of "sensori-motor intelligence". First the child's mind develops the representation of the world in terms of objects and actions, then the child learns to speak; and that initial speech (of one-word sentences) is "semantic", i.e. the initial relation between that representation of the world and sounds is purely semantic. As mental life evolves into more and more complex structures, so does language. Language acquisition is a process of hierarchic construction, and complexity of adult language is the result of that process. Chomsky's "universal grammar" is an illusion due to the fact that all children are programmed to develop through the same stages and achieve the same adult stage, and language simply reflects the outcome of that step-by-step hierarchical process.

The US developmental psychologist Elizabeth Bates pointed out that the development of language occurs while many other cognitive faculties are developing. She believes that language is not "one" isolated phenomenon but the result of a number of cognitive developments, each of which affects more than one cognitive faculty and the sum of which accounts for the development of all cognitive faculties, including language. In other word, there is no program for learning to speak, but there are several programs to learn several skills, which, together, enable "also" language. For example, we learn to play chess, but that does not mean that a program to play chess is present in our genetic information. Playing chess requires a number of skills, shared with many other tasks, that are enabled by our genetic information.

According to Bates, there is no "universal grammar" à la Chomsky. There is a global development of interconnected cognitive skills.

Adaptive Grammar

The US linguist Donald Loritz argued that rhythm is the "central organizing mechanism" of language. He shows that the sequence in which a child learns both phonology and morphology is based on the development of rhythms. Loritz shows that this phenomenon has a neural basis.

Loritz notes that children learn to walk before they learn to talk. Their learning of talking improves exponentially after they have learned to walk. He argues that walking introduces a "rhythmic dipole" into the child's brain, which has the effect of organizing the child's sounds: "babbling gets rhythm and becomes speech".

Rejecting Chomsky's generative grammar, Loritz wants to found "adaptive grammar", which is built around the "topic". Neurally speaking, the topic is the most active resonance. Loritz claims that discourse and sentences are built around the topic: "Syntax is ordered by topicality". This is also how grammar is grounded into reality: it is built around reality, a reality that is active in the brain in the form of a resonance.

Loritz argues that this phenomenon may recapitulate how language originated in the first place.

Further Reading
Allen, James: NATURAL LANGUAGE UNDERSTANDING (Benjamin Cummings, 1995)
Bach, Emmon: CATEGORIAL GRAMMARS (Reidel, 1988)
Baker, Mark: THE ATOMS OF LANGUAGE: THE MIND'S HIDDEN RULES OF GRAMMAR (Basic Books, 2001)
Bar-Hillel, Yehoshuas: LANGUAGE AND INFORMATION (Addison Wesley, 1964)
Bates, Elizabeth: THE EMERGENCE OF SYMBOLS (Academic Press, 1979)
Bresnan, Joan: MENTAL REPRESENTATIONS OF GRAMMATICAL RELATIONS (MIT Press, 1982)
Brown, Roger: A FIRST LANGUAGE (Harvard Univ Press, 1973)
Chierchia, Gennaro: MEANING AND GRAMMAR (MIT, 1990)
Chierchia, Gennaro: DYNAMICS OF MEANING (Univ of Chicago Press, 1995)
Chomsky, Noam: SYNTACTIC STRUCTURES (Mouton, 1957)
Chomsky, Noam: ASPECTS OF THE THEORY OF SYNTAX (MIT Press, 1965)
Chomsky, Noam: REFLECTIONS ON LANGUAGE (Pantheon, 1975)
Chomsky, Noam: THE LOGICAL STRUCTURE OF LINGUISTIC THEORY (University of Chicago Press, 1975)
Chomsky, Noam: RULES AND REPRESENTATIONS (Columbia Univ Press, 1980)
Chomsky, Noam: LECTURES ON GOVERNMENT AND BINDING (MIT Press, 1981)
Chomsky, Noam: KNOWLEDGE OF LANGUAGE (Greenwood, 1986)
Devlin, Keith: GOODBYE DESCARTES (John Wiley, 1997)
Dowty, David: WORD MEANING AND MONTAGUE GRAMMAR (Reidel, 1979)
Dowty, David: INTRODUCTION TO MONTAGUE SEMANTICS (Reidel, 1981)

Fauconnier, Gilles: MENTAL SPACES (MIT Press, 1994)
Fauconnier, Gilles & Eve Sweetser: SPACES, WORLDS, AND GRAMMAR (Univ of Chicago Press, 1996)
Gazdar, Gerald: GENERALIZED PHRASE STRUCTURE GRAMMAR (MIT Press, 1985)
Goddard, Cliff & Wierzbicka, Anna: SEMANTIC AND LEXICAL UNIVERSALS (Benjamins, 1994)
Jackendoff, Ray: SEMANTICS AND COGNITION (MIT Press, 1983)
Jackendoff, Ray: SEMANTIC STRUCTURES (MIT Press, 1990)
Jackendoff, Ray: LANGUAGES OF THE MIND (MIT Press, 1992)
Katz, Jerrold: AN INTEGRATED THEORY OF LINGUISTIC DESCRIPTIONS (MIT Press, 1964)
Katz, Jerrold: SEMANTIC THEORY (Harper & Row, 1972)
Katz, Jerrold: THE METAPHYSICS OF MEANING (MIT Press, 1990)
Korzybski, Alfred: SCIENCE AND SANITY - AN INTRODUCTION TO NON-ARISTOTELIAN SYSTEMS AND GENERAL SEMANTICS (1933)
Lakoff, George: PHILOSOPHY IN THE FLESH (Basic, 1998)
Lehnert, Wendy: STRATEGIES FOR NATURAL LANGUAGE LANGUAGE (Lawrence Erlbaum, 1982)
Langacker, Ronald: FOUNDATIONS OF COGNITIVE GRAMMAR (Stanford Univ Press, 1986)
Lenneberg, Eric: BIOLOGICAL FOUNDATIONS OF LANGUAGE (Wiley, 1967)
LePore, Ernest: NEW DIRECTIONS IN SEMANTICS (Academic Press, 1987)
Loritz, Donald: HOW THE BRAIN EVOLVED LANGUAGE (Oxford Univ Press, 1999)
Lycan, William: LOGICAL FORM IN NATURAL LANGUAGE (MIT Press, 1984)
Montague, Richard: FORMAL PHILOSOPHY (Yale University Press, 1974)
Nelson, Katherine: LANGUAGE IN COGNITIVE DEVELOPMENT (Cambridge University Press, 1996)
Pinker, Steven: THE LANGUAGE INSTINCT (William Morrow, 1994)
Pinker, Steven: HOW THE MIND WORKS (Norton, 1997)
Sapir, Edward: LANGUAGE (1921)
Schank, Roger: CONCEPTUAL INFORMATION PROCESSING (North Holland, 1975)
Trubetzkoy, Nicholas: PRINCIPLES OF PHONOLOGY (1939)

Van Benthem, Johan: A MANUAL OF INTENSIONAL LOGIC (Univ Of Chicago Press, 1988)
Van Benthem, Johan: LANGUAGE IN ACTION (MIT Press, 1995)
Vygotsky, Lev: THOUGHT AND LANGUAGE (MIT Press, 1934)
Whorf, Benjamin Lee: LANGUAGE, THOUGHT AND REALITY (MIT Press, 1956)
Wierzbicka, Anna: SEMANTICS, CULTURE, AND COGNITION (Oxford University Press, 1992)
Wierzbicka, Anna: THE SEMANTICS OF GRAMMAR (Benjamins, 1988)
Wierzbicka, Anna: SEMANTIC PRIMITIVES (1972)

The History of Language: Why We Speak

The Origin of Language

Charles Darwin observed that languages seem to evolve the same way that species evolve. However, just like with species, he failed to explain what the origin of language could be.

Languages indeed evolved just like species, through little "mistakes" that were introduced by each generation. It is not surprising that the evolutionary trees drawn by biologists (based on DNA similarity) and linguists (based on language similarity) are almost identical. Language may date back to the beginning of mankind.

What is puzzling, then, is not the evolution of modern languages from primordial languages: it is how it came to be that non-linguistic animals evolved into a linguistic animal such as the human being. The real issue is the "evolution of language" from non-language, not the "evolution of languages" from pre-existing languages, that is puzzling.

In fact, this is so puzzling that the US biologist Stephen Jay Gould, taking the view that Nature is more opportunistic than rational, held that language did not evolve from primitive forms of communication but by accident (or, better, by "exaptation") as a consequence of neural circuits that evolved for different functions. First thinking appeared and later language arose as one of its by-products.

A useful cue for research on the origin of language is that makes human language unique. The Indian neurologist Vilayanur Ramachandran thinks that there are five unique features of human language: an enormous lexicon; function words (not just sounds to signify an object or an event) such as "if ... then ..."; the capacity to refer to things that don't exist, or that existed in the past or that will exist in the future; the ability to create and understand metaphor and analogy; and its recursive structure (we can say and understand sentences such as "he thought that she thought...").

According to the US linguist Noam Chomsky and the US biologist Marc Hauser ("The faculty of language", 2002), what is unique to human language is only recursion.

There isn't yet a plausible explanation for why only humans evolved the sophisticated combinatorial language that they use. Perhaps this is because we are not looking in the right place. Something must have made human language not only necessary but indispensable, and that must be something that has to do with the kind of lives that humans live.

Humans are the only species that does not like to live the way previous generations lived. Children disobey, and teenagers are rebels. In theory, this makes no evolutionary sense: the children despise the knowledge accumulated by the parents. Other species don't exhibit this tendency: the children live the exact same kind of life that their parents lived. The lifestyle of a squirrel is the same as the lifestyle of a squirrel that lived thousands of years ago. On the other hand, the lifestyle of humans changes with every generation.

If there were no history (change in what we do and how we do it), there would be no need for the sophisticated human language. The minds of other species only need a language that communicates what is happening now.

Not only their language but even the "sentences" and the "words" of those "languages" have presumably remained the same throughout the centuries: there is nothing new to talk about.

On the other hand, minds whose lifestyle changes from one generation to the next one (i.e. human minds) need a much broader language, and, more importantly, a language that needs to be flexible: humans don't know what they will be speaking about tomorrow, so their language must already contain tomorrow's language as a possible language.

Minds that continuously change their world need a language that can continuously change, and need a language that is capable of referring to the past. If your world is not your father's world, it takes more than simple present-based language. It takes a language that can describe both your father's world and your world.

Humans need to be able to say things that have never been said before.

Animals only need to be able to repeat the same things that all previous generations have said.

Language as a By-product of Symbolization

The US philosopher Susanne Langer argued that the human mind differs from the minds of other animals in one key aspect. The mind of an animal is a transmission of stimuli from the world to the motor centers of the body. On the contrary, humans use signals as reminders. We can use signals to think of things that are not there. We can focus and discuss objects that are not present. The human mind deals with symbols, not just signals. The downside of symbols is that we are sometimes wrong. An animal that recognizes a signal rarely makes a mistake, because the environment checks its determination. A human who is using a symbol is more likely to make a mistake, because its determination is only checked by other human minds. The by-products of language do not complicate the world of an animal. Langer points out that ritual and magic are symbolic activities that, from an animal's point of view, are hopelessly senseless: an animal would never dance around a fire the way a man dances around a fire to make something happen. Animals have a direct relationship to events in their world. Humans construct huge symbolic universes that separate them from reality.

Langer concludes that the reason humans do such strange things with symbols must be that humans are symbolic systems at a biological level. They cannot escape the fate of creating and using symbols. It is a process that goes on all the time, whether consciously or unconsciously. It is built into the physical structure of the human brain. The brain continuously and endlessly builds symbols out of the sensory input. Since the sensory input never stops, the brain never stops building and interpreting symbols. Ideas pop up spontaneously. That is human nature. We just cannot help abstracting the world (i.e., thinking). Processing symbols serves a purpose but, at the same time, it constitutes an end in itself: we are programmed to process symbols.

By the same token, Langer thinks that ritual and magic are spontaneous activities, the by-products of the human mind's propensity for transforming everything into symbols. It is not a rational process, but an unconscious one. The propensity for symbolic processing can grow forever, even to the point that it becomes no longer useful and even harmful.

The evolutionary advantage of "conceiving" a thing as a symbol is that at the physical level no two people see the same thing (each brain is slightly different) but all people form the same symbol of the same thing. If we simply exchange a pixel map of what we saw, we would not find any two identical matches; but what we exchange is the concepts we formed of the respective pixel maps, and those are likely to be identical if the thing is the same.

Humans benefited from such exchanges of symbols. The reason that language became the primary form of communication is that, as Bertrand Russell originally noted, speech is the most economical way of rapidly producing many symbols via bodily movement. We could in theory use hand signals or facial expressions or shoulder movements, but it would be a lot more demanding from a physical point of view. Speech only requires movements of the lips and the tongue. Speech makes it very easy to combine many symbols into groups and therefore refer to situations (as opposed to individual concepts).

Langer points out that many "things" cannot be expressed in language but are still symbols. Language is not the only "language" we employ. It is just the most efficient. Langer sees ritual, myths and music as parallel non-linguistic languages: she calls it "presentational symbolism" instead of "discursive symbolism". They all arise form the brain's inescapable propensity to group sensory inputs into "forms" (as she refers to the Gestalt psychologists of the time). We recognize two situations as alike not because they provide identical sensory input but because they are analogous.

Following Edward Sapir, Langer thinks that language was not born to communicate. Communication is a by-product of symbolization. Our brains create symbols all the time, whether we want to communicate them or not, and it turns out that symbols constitute a very effective way to communicate. Therefore we started using language to communicate, but language pre-existed linguistic communication. Originally language was only for "naming" things, or, better, concepts. Because of its nature, though, it was inevitable for language to evolve into a communication medium in every human civilization. Langer notes that babies tend to babble spontaneously, whereas other primates don't. From the very beginning the child wants to transform her experience into vocal sounds. Langer thinks that this proves how language's mission is to transform experience into symbols (concepts). In a sense, it is not true that a child has to learn to speak. A child has to learn to speak the language of the parents, but the child is already speaking from the very first moment of life. A child uses sounds all the time to complement all her experiences. The set of those sounds is consistent and (as far as the horizon of that child goes) complete, and therefore constitutes a language, albeit a language that only that child can understand. Her parents teach the child a language that is to be shared with the community. They don't teach the child to speak: they teach the child to speak a specific language.

Following the Danish linguist Otto Jespersen, Langer argued that singing and dancing came first. The languages of primitive cultures have a singsong quality that has been lost in modern languages, but most likely all languages originally were "sung". The language of children fluctuates violently in tone. As "civilized" adults, we still use fluctuations in our tone in order to deliver the real meaning of sentences, but the fluctuations are vastly downplayed. Somehow we decided that those fluctuations were not "polite". Singing and speaking became two different things; and today we teach children not to scream, not to cry, not to jubilate, and so forth, thus progressively eliminating the "singing" quality of language. Proper erudite talk strives to remove all fluctuations.

Langer did not speculate on the reason that induced us to weed out fluctuations/emphases from speech, a fact that is not obvious since those fluctuations in tone help deliver the meaning: why do we speak instead of singing?

One also wonders where rhythm comes from. Most likely from nature itself, from the countless biorhythms that regulate the body and that ultimately reflect the countless rhythms of nature; and from the rhythms of work in the fields.

Langer argues that music was not born as an art but as a combination of symbolic activities: dance and song. Following the German philosopher Wilhelm von Humboldt, she thinks that humans are "singing animals". We naturally sing, not for artistic purposes but simply because it's in our nature. Following the Swiss music theorist Ernst Kurth, she thinks that music has accumulated "ursymbols" (primordial symbols) that refer to familiar sounds in nature, industry or society. By the same token, we naturally dance. Singing and dancing predate music as an art form.

While Langer did not want to ascribe any symbolic capability to other animals, her theory could also explain how bees communicate by dancing and birds communicate by singing.

Langer viewed all forms of symbolization as originating from the same principle: the mind creates symbols all the time, and then some of its symbolic activity turns out to be important for some practical activity. We often confuse the importance and the origin of a phenomenon. We think that speech is for communicating because language is important for communicating, when in reality communication was just a by-product of speech. By the same token, music was born for non-artistic reasons, as yet another manifestation of the mind's endless activity of symbol processing, but then music became important as an "art" to express feelings that couldn't be expressed by speech alone.

An Evolutionary Accident

Several biologists and anthropologists believe that language was "enabled" by accidental evolution of parts of the brain and possibly other organs.

The US biologist Philip Lieberman views the brain as the result of evolutionary improvements that progressively enabled new faculties. Human language is a relatively recent evolutionary innovation that came about when speech and syntax were added to older communication systems. Speech allowed humans to overcome the limitations of the mammalian auditory system, and syntax allowed them to overcome the limits of memory.

The US neurologist Frank Wilson believes that the evolution of the human hand allowed humans a broad range of new activities that, in turn, fostered an evolution of the brain that resulted in the brain of modern humans. Anatomical changes of the hand dramatically altered the function of the hand, eventually enabling it to handle and use objects. This new range of possibilities for the hand created a new range of possibilities for thought: the brain could think new thoughts and could structure them. The human brain (and only the human brain) organizes words into sentences, i.e. does syntax, because of the hand. "The brain does not live inside the head".

Linguistic Darwinism

According to Chomsky's classical theory, language is an innate skill: we come pre-wired for language, and simply "tune" that skill to the language that is spoken around us. In Chomsky's view language is biology, not culture. This implies that the language skill is a fantastic byproduct of evolution. Syntax must be regarded as any other organ acquired via natural selection. How did such a skill develop, since that skill is not present elsewhere in nature? Where did it come from? Language appears to be far too complex a skill to have been acquired via step-by-step refinement of the Darwinian kind, especially since we are not aware of any intermediary steps (e.g., species that use a grammar only to some extent).

The British linguist Derek Bickerton advanced a theory that attempted to bridge Darwin and Chomsky. Bickerton argued that language was the key to the success of the human species, the one feature that made us so much more powerful than all other species. Everything else, from memory to consciousness, seems to be secondary to it. We cannot recall any event before we learned language. We can remember thoughts only after we learned language. Language seems to be a precondition to all the other features that we rank as unique to humans.

First of all, human language cannot just be due to the evolution of primitive, emotion-laden "call systems". We still cry, scream, laugh, swear, etc. Language has not fully replaced that system of communication. The primitive system of communication continues to thrive alongside language. Language did not replace it, and probably did not evolve from it. Language is something altogether different.

He emphasized the difference (not the similarity) between human and animal communication. Animal communication is holistic: it communicates the whole situation. Human language deals with the components of the situation. Furthermore, animal communication is pretty much limited to what is evolutionarily relevant to the species. Humans, on the other hand, can communicate about things that have no relevance at all for our survival. In fact, we could adapt our language to describe a new world that we have never encountered before. The combinatorial power of human language is what makes it unique. Bickerton thinks that human and animal communication are completely different phenomena.

In fact, Bickerton believes that human language is not primarily a means to communicate but a means to represent the world. Human language did not evolve from animal communication but from older representation systems. First, some cells (the sensory cells) were born whose only task was to respond to the environment. As sensory cells evolved and their inputs became more complex, a new kind of cells appeared that was in charge of mediating between these cells and motor cells. These mediating cells eventually evolved categories that were relevant to their survival. Animals evolved that were equipped with such "primary" representational systems. At some point, humans evolved who were equipped with syntax and were capable of representing representations (of models of models). Human language was so advantageous that it drove a phenomenal growth in brain size (not the other way around).

Two aspects of language, in particular, set it apart from the primitive call system of most animals: the symbolic and the syntactic aspects. A word stands for something (such as an object, a concept, an action). And words can be combined to mean more than their sum ("I walk home" means more than just the concepts of "i", "walking" and "home"). Bickerton believes that syntax is what makes our species unique: other species can also "symbolize", but none has showed a hint of grammar.

The philosopher Nicholas Humphrey once advocated that language was born out of the need to socialize. On the contrary, Bickerton believes that Humphrey's "social intelligence" had little to do with the birth of proto-language. Socialization as a selective pressure would not have been unique to humans, and therefore language would have developed as well in other primates. Syntax, instead, developed only in humans, which means that a selective pressure unique to humans must have caused it. Bickerton travels back to the origins of hominids, to the hostile savannas where hominids were easy targets for predators and had precious little food sources. Other primates had a much easier life in the forests. The ecology of early hominids created completely different selective pressures than the ones faced by other primates. In his quest for the very first utterances, Bickerton speculates that language was born to label things, then evolved to qualify those labels in the present situation: "leopard footprints" and "danger" somehow needed to be combined to yield the meaning "when you see leopard footprints, be careful".

Bickerton shows how this kind of "social calculus", coupled with Baldwin effects, could trigger and successfully lead to the emergence of syntax. Social intelligence was therefore important for the emergence of syntax, even if it was not important for the emergence of proto-language.

Bickerton points out that the emergence of language requires the ability to model other minds. I am motivated to communicate information only if I can articulate this simple scenario in my mind: I know something that you don't know and I would gain something if you knew it. Otherwise, the whole point of language disappears.

Bickerton thinks that consciousness and the self were enabled by language: language liberated humans from the constraints of animal life and enabled off-line thinking. The emergence of language even created the brain regions that are essential to conscious life. Basically, he thinks that language created the human species and the world that humans see.

To summarize, Bickerton believes that: language is a form of representation, not just of communication, a fact that sets it apart from animal communication; language evolved from primordial representational systems; language has shaped the cognitive life of the human species.

Grooming

The British psychologist Robin Dunbar believes that originally the function of language was not to communicate information, but to cement society. From the point of view of information, language brings more benefits to the listener than to the talker. If that were the main purpose of language, it would have caused the evolution of a race of listeners, not of talkers, and far less of gossipers.

All primates live in groups. The size of a primate's neocortex, as compared to the body mass, is directly proportionate to the size of the average social group for that primate. Humans tend to live in the larger societies of primates, and human brains are correspondingly much larger.

Dunbar studied how brain size determines social size and why. Every species has a cognitive limit to the number of individuals with whom any one individual can maintain stable relationships. That number for humans is about 150. Generally speaking, the bigger the brain (the neocortex) the higher the number of relationships among objects that the brain can handle. Each species has a natural limit to how much its brain can "correlate".

As humans transitioned from the forest to the savanna, they needed to band together in order to survive the increased danger of being killed by predators. Language helped keep large groups together. Thus humans who spoke had an evolutionary advantage (the group) over humans who did not develop that skill. Dunbar believes that human speech is simply a more efficient way of "grooming": apes cement social bonds by grooming the members of their group. Humans "gossiped" instead of grooming each other. Later, and only later, humans began to use language also to communicate information.

Dunbar believes that dialects developed for a similar reason: to rapidly identify members of the same group (it is notoriously difficult to imitate a dialect).

Language and societies evolved together: society stimulated the invention of language, and language enabled larger societies, that stimulated even more sophisticated languages, that enabled even larger societies, etc.

Brain size is also part of the equation, though. Basically, in order to avoid being eaten by predators, humans needed to band together, and that required bigger brains.

Co-evolution

The US anthropologist Ralph Holloway ("The evolution of the human brain", 1967) claimed already in the 1960s that brain reorganization must have preceded brain evolution: ancient hominids still had a small brain like apes but could already do things that an ape's brain couldn't do like full bipedalism. He posited a form of positive feedback between "mental" evolution and bodily evolution.

The US anthropologist Terrence Deacon believes that language and the brain co-evolved. They evolved together influencing each other step by step. In his opinion, language did not require the emergence of a language organ. Language originated from symbolic thinking, an innovation that occurred when humans became hunters because of the need to overcome the sexual bonding in favor of group cooperation.

Both the brain and language evolved at the same time through a series of exchanges. Languages are easy to learn for infants not because infants can use innate knowledge but because language evolved to accommodate the limits of immature brains. At the same time, brains evolved under the influence of language through the Baldwin effect. Language caused a reorganization of the brain, whose effects were vocal control, laughter and sobbing, schizophrenia, autism.

Deacon rejects the idea of a universal grammar à la Chomsky. There is no innate linguistic knowledge. There is an innate human predisposition to language, but it is due to the co-evolution of brain and language and it is altogether different from the universal grammar envisioned by Chomsky. What is innate is a set of mental skills (ultimately, brain organs) which translate into natural tendencies, which translate into some universal structures of language.

Another way to describe this is to view language as a "meme". Language is simply one of the many "memes" that invade our mind. And, because of the way the brain is, the meme of language can only assume such and such a structure: not because the brain is pre-wired to such a structure but because that structure is the most natural for the organs of the brain (such as short-term memory and attention) that are affected by it.

Chomsky's universal grammar is an outcome of the evolution of language in our mind during our childhood. There is no universal grammar in our genes, or, better, there are no language genes in our genome.

The secret of language is not in the grammar, but in the semantics. Language is meaningful. Deacon envisions a hierarchy of levels of reference (of meaning), that reflects the evolution of language. At the top is the level of symbolic reference, a stable network of interconnected concepts. A symbol does not only refer to the world, but also to other symbols. The individual symbol is meaningless: what has meaning is the symbol within the vast and ever changing semantic space of all other symbols. At lower levels, Deacon envisions less and less symbolic forms of representation, which are also less and less stable. At the lowest, most fluctuating level of the hierarchy there lie references that are purely iconic and indexical, created by a form of learning that is not unique to language (in fact, it is widespread in all cognitive tasks). The lower levels are constrained by what humans can experience and learn, which is constrained by innate abilities. The higher level, on the other hand, is an emergent system due to the interaction among linguistic agents.

Gesturing in the Mind

According to US neuroscientist Rhawn Joseph, one of the youngest parts of the brain, the inferior parietal lobe of the left hemisphere, enabled both language, tool making and art itself. It enabled us, in other words, to create visual symbols. It also enabled us to create verbal symbols, i.e. of writing.

The inferior parietal lobe allows the brain to classify and label things. This is the prerequisite to forming concepts and to "abstracting" in general. Surprisingly, this is also the same organ that enables meaningful manual gesturing (a universal language, that it is also shared with many animals). Thus the evolution of writing is somehow related (neurally speaking) to manual gesturing. The inferior parietal lobe was one of the last organs of the brain to evolve, and it is still one of the last organs to mature in the child (which explains why children have to wait for a few years before they can write and do math).

This lobe is much more developed in humans than in other animals (and non-existent in most). The neurons of this lobe are somewhat unique in that they are "multimodal": they are capable of simultaneously processing different kinds of inputs (visual, auditory, movement, etc). They are also massively connected to the neocortex, precisely to three key regions for visual, auditory and somatosensory processing. Their structure and location makes them uniquely fit to handle and create multiple associations. It is probably this lobe that enables us to understand a word as both an image, a function, a name and many other things at the same time.

Joseph claims that the emotional aspect of speaking is the original one: the motivation to speak comes from the limbic system, the archaic part of the brain that deals with emotions, and that we share with other mammals. The limbic system embodies a universal language that we all understand, a primitive language made of calls and cries. Each species has its own, but within a species all members understand it. Joseph believes that at this stage the "vocal" hemisphere is the right one. Only later, after a few months, does the left hemisphere impose structure to the vocalizing and thus become dominant in language.

Language as A Sexual Organ
The US evolutionary psychologist Geoffrey Miller believes that the human mind was largely molded by sexual selection and is therefore mainly a sexual ornament. Culture, in general, and language, in particular, are simply ways for males and females to play the game of sex. When language appeared, it quickly became a key tool in sexual selection, and therefore it evolved quickly.

Darwin had already speculated that language may have evolved through sexual selection(as a means to impress sexual mates). Miller agrees, finding that the usual explanation (that language helps a group trade key information) is only a small piece of the puzzle (individuals, unless they are relatives, have no motivation to share key information since they are supposed to compete).

Even more powerful is the evidence that comes from observing the behavior of today's humans: they compete to be heard, they compete to utter the most sensational sentences, they are dying to talk.

Miller also mentions anatomical evidence: what has evolved dramatically in the human brain is not the hearing apparatus but the speaking apparatus. Miller believes that language, whose intended or unintended effect is to deliver knowledge to competitors, must also have a selfish function, otherwise it would not have developed: individuals who simply delivered knowledge to competitors would not have survived. On the other hand, if language is a form of sexual display, then it makes sense that it evolved rapidly, just like any other organ (bull horns or peacock tails) that served that function. It is unique to humans the same way that the peacock's tail is unique to peacocks. It is pointless to try and teach language to a chimpanzee the same way that it is pointless to expect a child to grow a colorful tail.

The Origin of Communication
Where does language come from is a question that does not only apply to humans, but to all species, each species having its own "language".

One might as well ask the question, "Where do bee dances come from"? The bees are extremely good at providing details about the route and the location of food. They do so not with words but by dancing. The origins of bee dances are no less intriguing than the origins of human language.

The point is that most species develop a social life and the social life depends on a mechanism of communication, and in humans that mechanism is language. But language may be viewed as a particular case of a more general process of nature, the process by which several individuals become aggregated in a group.

There is a bond within the members of a species, regardless of whether they are cooperating or competing: they can communicate. A dog cannot communicate much to a cat. A lion cannot communicate with an ant. And the greatest expert in bees cannot communicate much with a bee. Communication between members of different species is close to impossible. But communication within members of a species is simple, immediate, natural, and, contrary to our beliefs, does not require any advanced skills. All birds communicate; all bees communicate. There is no reason to believe that humans would not communicate if they were not taught a specific language. They might, in fact, communicate better: hand gestures and facial expressions may be a more efficient means of communication among humans than words.

Again, this efficiency is independent of the motives: whether it is for cooperation or for aggression. We can communicate with other members of our species. When we communicate for cooperation, the communication can become very complex and sophisticated. We may communicate that a herd is moving east, that clouds are bringing rain, that the plains are flooded. A bee can communicate similar information to another bee. But an ant cannot communicate this kind of information to a fish and a fish cannot communicate it to a bird. Each species has developed a species-specific form of communication.

The origin of language is but a detail in a much more complex story, the story of how intra-species communication evolved. If all species come from a common ancestor, there must have been only one form of communication at the beginning. Among the many traits that evolved over the ages, intra-species communication is one that took the wildest turns. While the genetic repertoire of bees and flies may be very similar, their system of communication is quite different.

The fact that communication is different for each species may simply be due to the fact that each species has different kinds of senses, and communication has to be tailored to the available senses.

A reason for this social trait to exist could be both sexual reproduction and altruism.

The Origin of Cellular Communication

Even before social behavior was invented, there was a fundamental language of life. Living cells communicate all the time, even in the most primitive organisms: cell communication is the very essence of being alive.

There are obvious parallels between the language of words and the language of cellular chemicals. Two cells that exchange chemicals are doing just that: "talking" to each other, using chemicals instead of words. Those chemicals are bound in molecular structures just like the words of human language are bound in grammatical structures.

The forms of communication that do not involve chemical exchange still cause some chemical reaction. A bee that changes course or a human brain that learns something have undergone chemical change, that has triggered changes in their cognitive state.

From this point of view, there are at least three main levels of communication: a cellular level, in which living cells transmit information via chemical agents; a bodily level, in which living beings transmit information via "gestures"; and a verbal level, in which living beings transmit information via words.

Each level might simply be an evolution of the previous one.

At the same time, the Norwegian mathematician Nils Barricelli ("The Functioning of Intelligence Mechanisms Directing Biologic Evolution," 1985) warned that the kind of communication occurring at the cellular level is different from the kind of communication occurring among animals: "If humans, instead of transmitting to each other reprints and complicated explanations, developed the habit of transmitting computer programs allowing a computer-directed factory to construct the machine needed for a particular purpose, that would be the closest analogue to the communication methods among cells of various species."

Who Invented Language?

Linguists, geneticists and anthropologists have explored the genealogical tree of human languages to determine where human language was invented. Was it invented in one place and then spread around the globe (why then so many languages rather than just one?) or was it invented in different places around the same time? (What a coincidence that would be).

The meta-issue with this quest is the role of free will, i.e. whether we humans have free will and decide what happens to us. We often assume that somebody "invented" something and then everybody started using it. The truth could be humbler: all humans share pretty much the same brain, and that brain determines our behavior. We all sleep, we all care for our children, and we all avoid danger. Not because one human "invented" these behaviors, but because our brains are programmed to direct us to behave that way. Our free will (if indeed we have any) is limited to deciding which woman to marry, but the reason we want a wife is sex and children, a need that is programmed in our brain (and, of course, one could claim that the choice of the specific wife is also driven by our brain's circuits).

In fact, we consider "sick" or "abnormal" any human being who does not love her/his children, any human who does not like sex, etc.

Asking who invented language could be like asking who invented sex or parenting. It may just come with the race. We humans may be programmed to communicate using the human language. It didn't take a genius to invent language. We started speaking, worldwide, as soon as the conditions were there (as soon as we started living in groups, more and more heterogeneous groups, more and more collaborative groups).

The mystery may not be who invented language, but why we invented so many and so different languages. There are striking differences between Finnish and Chinese, even though those two peoples share pretty much the same brain. The effect of the environment on the specific language we start speaking must indeed be phenomenal.

What Are Jokes And Why Do We Make Them

Language developed because it had an evolutionary function. In other words, it helped us survive. For example, language enabled humans to exchange information about the environment. A member of a group can warn the member of another group about an impending danger or the source of water or the route taken by a predator.

This may be true, but it hardly explains the way we use language every day. When we write an essay, we may be matter of factual, but most of the day we are not. For example, we make jokes all the time. A human being who does not make jokes, or does not laugh at jokes made by others, is considered a case for a psychoanalyst. Jokes are an essential part of the use of language.

Nonetheless, jokes are a peculiar way to use language. We use words to express something that is not true, but could be true, and the brain somehow relates to this inconsistency and... we laugh.

There must be a reason why humans make jokes. There must be a reason why we use language to make jokes.

Upon closer inspection, we may not be so sure that the main function of language is communicating information about the environment.

If a tiger attacks you, i will not read you an essay on survival of the fittest: i will just scream, "Run!" We don't need the complex, sophisticated structure of language to "communicate" about the environment and us. If you are starving, I may just point to the refrigerator. For most practical purposes, street signs communicate information about locations better than geography books. It is at least debatable whether we need language to communicate information about the environment that is relevant to survival. We can express most or all of that information in very simple formats, often with just one word or even just a gesture.

On the other hand, if we want to make a joke, we need to master the whole power of the language. Every beginner in a foreign language knows that the hardest part is to understand jokes in that language, and the second hardest is making them. Joking does require the whole complex structure of language, and, at closer inspection, it is the only feature of human life that requires it. In fact, it requires even more than that. Making (and understanding) jokes is possible only if you have a full understanding of the context. A computer that wants to "laugh" or make someone laugh will have to master much more than just the linguistic kills.

Jokes are probably very important for our survival. A joke is a practice: we laugh because we realize that something terrible would happen in that circumstance: the logic of the world would be violated, or a practical disaster would occur. The situation is "funny" because it has to be avoided. Being funny helps remember that we should avoid it.

Joking may well be an important way to learn how to move in the environment without having to do it first person, without having to pay the consequences for a mistake.

In that case, it would be more than justified that our brain evolved a very sophisticated tool to make jokes: language.

Ultimately, language may have evolved to allow us to make more and more useful (funnier and funnier) jokes.

Tools

The British psychologist Richard Gregory has shown how language is but one particular type of "tool". The human race, in general, is capable of making and using tools, and language happens to be one of them.

Gregory claims that "tools are extensions of the limbs, the senses and mind." The fundamental difference between humans and apes is not in the (very small) anatomical differences but in language and tools. Man is both a tool-user and a tool-maker.

Gregory shows that there are "hand" tools (such as level, pick, axe, wheel, etc) and "mind" tools, which help measuring, calculating and thinking (such as language, writing, counting, computers, clocks).

Tools are extensions of the body. They help us perform actions that would be difficult for our arms and legs. Tools are also extensions of the mind. Writing extended our memory. We can make a note of something.

So do photographs and recordings. This book extends my mind. It also extends your mind. Tools like books create a shared mind.

Gregory qualifies information as "potential intelligence" and behavior as "kinetic intelligence". Tools increase intelligence as they enable a new class of behavior. A tool "confers" intelligence to a user, meaning that it turns some potential intelligence into kinetic intelligence.

A person with a tool is a person with a potential intelligence to perform an action that without the tool would not be possible (or much more difficult).

Behavior is often just using that tool to perform that action. It may appear that intelligence is in your action, but, actually, intelligence is in the tool, not in your action. Or, better, they are two different types of intelligence.

And words are just one particular type of tool.

There is also a physical connection in our body between language and tool usage: the same hemisphere controls them both.

Tools as Intentionality

The US philosopher Daniel Dennett advanced a theory of language based on his theory of "intentionality" (the ability to refer to something). Basically, his idea is that different levels of intentionality correspond to different "kinds" of minds.

The "intentional stance" is the strategy of interpreting the behavior of something (a living or non-living thing) as if it were a rational agent whose actions are determined by its beliefs and desires. This is the stance that we adopt, for example, when dealing with ourselves and other humans: we assume that we and the others are rational agents whose actions are determined by our beliefs and desires. Intentional systems are those to which the intentional stance can be applied, and they include artifacts such as thermostats and computers, as well as all living beings. For example, we can say that "This computer program wants me to input my name," or that "The tree bends south because it needs more light," (both "wants" and "needs" express desire).

The intentional stance makes the assumption that an intentional system has goals that it wants to achieve; that it uses its own beliefs to achieve its own goals, and that it is smart enough to use the right ones in the appropriate way.

It seems obvious that artifacts possess only "derived" intentionality, i.e. intentionality that was bestowed on them by their creators. A thermostat measures temperature because that is what the engineer designed it for. The same argument, though, applies to us: we are artifacts of nature and nature bestows on us intentionality. (The process of evolution created our minds to survive in an environment, which means that our mind is about the environment).

Dennett speculates that brains evolved from the slow internal communication systems of "sensitive" but not "sentient" beings when they became equipped with a much swifter communication agent (the electro-chemicals of neurotransmitters) in a much swifter communication medium (nerve fibers). Control was originally distributed around the organism in order to be able to react faster to external stimuli. The advent of fast electro-chemicals allowed control to become centralized, because now signals traveled at the speed of electricity. This also allowed control to become much more complex, as many more things could be done in a second.

"Evolution embodies information in every part of every organism". And that information is about the environment. A chameleon's skin, a bird's wings, and so forth, they all embody information about the medium in which their bodies live. This information does not need to be replicated in the brain as well. The organ already "knows" how to behave in the environment. Wisdom is not only in the brain; wisdom is also embodied in the rest of the body. Dennett speculates that this "distributed wisdom" was not enough: a brain can supplement the crudeness, the slowness, and the limitations of the organs. A brain can analyze the environment on a broader scale, can control movement in a much faster way and can predict behavior over a longer range.

As George Miller put it, animals are "informavores". Dennett believes in a distributed information-sucking system, each component of which is constantly fishing for information in the environment. They are all intentional systems, which get organized in a higher-level intentional system, with an "increasing power to produce future".

This idea, both evolutionarily and conceptually, can be expressed in a number of steps of intentionality, each of which yields a different kind of mind. First there were "Darwinian creatures", that were simply selected by trial and error on the merits of their bodies' ability to survive (all living organisms are Darwinian creatures). Then came "Skinnerian creatures", which were also capable of independent action and therefore could enhance their chances of survival by finding the best action (they are capable of learning from trial and error). The third stage of "mind, "Popperian creatures", were able to play an action internally in a simulated environment before they performed it in the real environment and could therefore reduce the chances of negative outcomes (information about the environment supplemented conditioning). Popperian creatures include most mammals and birds. They feel pain, but do not suffer, because they lack the ability to reflect on their sensations.

Humans are also "Gregorian creatures", capable of creating tools, and, in particular, of mastering the tool of language. Gregorian creatures benefit from technologies invented by other Gregorian creatures and transmitted by cultural heritage, unlike Popperian creatures that benefit only from what has been transmitted by genetic inheritance.

A key step in the evolution of "minds" was the transition from beings capable of an intentional stance towards others to beings capable of an intentional stance towards an intentional stance. A first-order intentional system is only capable of an intentional stance towards others. A second-order intentional system is also capable of an intentional stance towards an intentional stance. It has beliefs as well as desires about beliefs and desires. And so forth. Higher-order intentional systems are capable of thoughts such as "I want you to believe that I know that you desire a vacation".

This was not yet conscious life because there are examples, both among humans and among other animals, of unaware higher-order intentionality. For example, animals cheat on each other all the time, and cheating is possible only if you are capable of dealing with the other animal's intentional state (with the other animal's desires and beliefs), but Dennett does not think that animals are necessarily conscious. In other words, he thinks that one can be a psychologist without being a conscious being.

Dennett claims that our greater "intelligence" is due not to a larger brain but to the ability to "off load" as much as possible of our cognitive tasks into the environment. We construct "peripheral devices" in the environment to which those tasks can be delegated. We can do this because we are intentional: we can point to those things that we left in the environment. In this way the limitations of the brain do not matter anymore, as we have a potentially infinite area of cognitive processing. Most species rely on natural landmarks to find their way around and track food sources. But some species (at least us) have developed the skills to "create" their own landmark, and they can therefore store food for future use. They are capable of "labeling" the world that they inhabit. Individuals of those species alter the environment and then the altered environment alters their behavior. They create a loop to their advantage. They program the environment to program them.

Species that store and use signs in the environment have an evolutionary advantage because they can "off-load" processing. It is like "taking a note" that we can look up later, so we don't forget something. If you do not take a note, you may forget the whole thing.

Thanks to these artifacts, our mind can extend out into the environment. For example, the notepad becomes an extension to my memory.

These artifacts shape our environment. Our brains are semiotic devices that contain pointers and indices to the external world.

Tools Created the Brain

An even more radical theory of how important tools have been for human evolution was advanced by the British archeologist Timothy Taylor.

Humans are the weakest of the great apes. We cannot survive without clothes and houses. It seems irrational that the one ape that is so vulnerable ended up dominating every other species. In fact, humans should not have survived evolution at all because reproduction is so dangerous, complicated and (in the past) lethal (for both baby and mother), and then because children require so much attention and dedication lest they die of the silliest causes (compare with the cubs of other mammals that are self-sufficient within weeks if not days).

Taylor argues that technology is the solution to this apparent contradiction. Evolution is not only biological. The cultural-technological component is equally important, and it vastly favored humans over any other species. Technology is very much part of what we mean by "being human". Darwin's theory of evolution needs to be complemented with a story of how technology allowed humans to violate the very rules that Darwin found embedded in all other species. According to Darwin's blind algorithm of natural selection, humans should have gone extinct very quickly. Biological evolution (or, better, biological accident) accounts for humans developing the upright posture. That posture freed the hands, and allowed humans to make tools. That situation completely altered the normal course of biological evolution because, from that point on, technology introduced a parallel (non blind) algorithm. Humans started evolving not based on biological laws of evolution but based on technological factors. Taylor claims that "technology evolved us".

Taylor points out that humans seem to go against evolution: we are "biologically reduced". Our bodies have weakened since the Stone Age and keep weakening. We are less strong, less agile, less fast, etc than our prehistoric ancestors were. Something similar happens to wild animals when they get domesticated. By analogy, Taylor calls "self-domestication" the process by which humans got weaker while becoming more dependent on technology, i.e. while developing a smarter brain. This is another strange loop: it could be that brains got smarter as bodies got weaker, or it could be that bodies got weaker as brains got smarter. In fact, it is not even true that human brains got larger since prehistory: the human brain has shrunk by about 10% over the last ten thousand years. The Neanderthal man was not only stronger than Homo Sapiens, it also had a bigger brain and perhaps was in many ways smarter. But it was Homo Sapiens that survived, and went on to rule the planet. Taylor calls it "survival of the weakest". Taylor is convinced that tools, the great invention of Homo Sapiens, dramatically altered the terms of evolution.

This, however, leads to a vicious loop. Technology requires a smart brain to design it and build it; but a smart brain requires a lot of proteins which come from hunting, fire and cooking, which are made possible by technology. Therefore technology requires a brain that is made possible by technology. It is a "chicken and egg" kind of problem: one cannot exist without the other, but one must have come first.

Taylor argues that humans did not evolve the ability to make tools (which requires big brains which require tools) but tools caused humans to evolve that way. Chimps don't have big brains because they need the skull configured in such a way to allow for their powerful jaws. Having fire and tools, humans did not need those powerful features of the head and therefore lost them and therefore space was left over for a big brain to grow. He points out that there is a gap beween the date of the oldest chipped stone tool (2.5 million years ago) and the first hominids (2.3 million): 200 thousand years, a very long period of time.

However, there is a problem with big brains: they should be physically impossible. As a consequence of standing upright, bipedal beings have a smaller pelvis, and therefore a larger head for babies makes no sense. The chances of miscarriage increase dramatically. Therefore the babies with large heads should have been eliminated by natural selection. For a while our bipedal ancestors continued to have small heads, as one would expect, but then suddenly hominids began to develop large heads, and that sounds like a physical impossibility.

It gets worse. Human babies are incapable of walking upright for a long time. They are in fact incapable of doing most of the things required to survive. In a sense, all human babies are born prematurely, they are extra-uterine foetuses. In fact, the brain of human babies keeps developing at lightning speed during the first year (as opposed to the brain of the chimp, that is pretty much done and ready to go at birth).

Taylor concludes that something was needed to make all of this not only possible but inevitable. Females were the first tool-makers: the need for tools to carry plant food predates the need for hunting tools; and the need for carrying their babies around probably predates both. Bipedalism created the need for carrying babies and for carrying goods, a need that other apes solve by moving on all four. Women were constantly on the move and there were no kindergartens: they had to carry their babies with them all the time. In every single culture women have solved this problem in the same way: every single culture developed tools to carry babies (from a simple sling to the stroller). Hence the first tools were probably made by women who were the ones who had the mother of all problems to solve: protect your babies. Those first tools triggered the expansion of the human brain.

The brain of human babies is still so plastic that it can absorb whatever technology is available in a way that no other species can. Other species are condemned to use the brain they get at birth, whereas human children can adapt their brain to the civilization they find. Once invented, a tool can last forever, passed from one generation to the next simply by exposing the new-born babies to it.

Brain size started increasing, according to Taylor, after technology happened. And bodies started getting weaker because technology made it unnecessary to be strong and big. Taylor speculates that the rapid growth of the human brain was due to competition for technological supremacy. And it all started with the baby-carrying sling.

It is not intelligence that gave us tools because the earliest tools predate the rapid expansion of the brain that led to modern hominids. Brains began to grow in size after (not before) the first tools were invented. Tools made possible larger brains. Tools made possible intelligence.

Taylor also advances a theory about the first symbolic art: he thinks that ancient statuettes were the equivalent of today's mannequins in department stores (they were meant to exhibit clothes); and he thinks that cave paintings were meant to express the belief in divine creation, a rational consequence of knowing that someone makes something therefore someone must have made us. Whatever the original purpose, the beginning of symbolic art (40000 years ago) also marks the moment when human brains started "outsourcing" intelligence to our tools. Taylor claims that it also accounts for the reason that human brains have stopped expanding since that age: the need for brains decreases as technology gets smarter.

Technology has become the main driver of human evolution. We are our technology. Humans increasingly depend on technology. Technological change is accelerating. Eventually machines might get so smart that humans will not be able to comprehend them anymore. Ray Kurzweil termed it the "singularity". That singularity will allow humans to extend not only human intelligence but also the life expectancy of humans because it will overcome the material constraint of the human body in the form of artificial intelligences and avatars. Whether we will reach that singularity or not, Kurzweil starts the process leading to it from very recently. Taylor does something similar but places it at the very beginning of the evolution of Homo Sapiens: there would be no Homo Sapiens without that symbiosis between body and tools, between biology and technology. The life expectancy of Homo Sapiens would have been virtually zero because of the material limitations of the human body. The futuristic visions of people like Ray Kurzweil are actually naive: they assume that this will happen in the future, when in fact it has already happened, and it has always been the case. Both mentally and physically humans have been shaped by technology.

Exaptation

Evidence from brain lesions led the Indian neurologist Vilayanur Ramachandran to think that not only the brain has specialized circuits for language, but that there are probably three different specialized areas for three different tasks: lexicon, syntax and semantics. A patient with Broca's aphasia wants to convey meaning but has difficulty composing sentences. On the other hand, a patient with Wernicke's aphasia utters sentences that are well-formed, grammatically correct, but meaningless (they are also in a simplified syntax that shuns recursion). Therefore it appears that Broca's area is specialized for syntax, and Wernicke's area is specialized for semantics.

Like Stephen Jay Gould, Ramachandran too thinks that language evolved as a by-product of other features, but not quite of thinking. He suspects that language emerged out of interference inside the brain between signals that are similar. The sound that refers to an object is similar (in some physical manner) to the image of that object that is being processed by the brain. He calls this "the synesthetic theory". The brain might "translate" an image into a sound (into a protoword) simply because the maps that represent the two are adjacent and interfere with each other in quite a natural fashion. This cross-activation would be the primal cause of our linguistic competence. Next, Broca's area indirectly relates syntax and the movements that we use to produce speech. Hence one part of the brain puts in contact image and sound, and then another part of the brain causes those sounds to be spoken. Each of these correlations can be view as an "abstraction", which in physical terms simply means that different brain regions activate each other based on similarity of signals. A third case of cross-activation exists. Darwin himself noted that sometimes an utterance is accompanying with a gesture. Ramachandran speculates that it could be another case of cross-activation, except this one would be due to the interaction between two motor maps. This would explain how a primordial language of gestures would evolve into a spoken language. A brain that exhibits the kind of wiring that the human brain has would naturally end up translating gestures into (spoken) words. "Abstraction" is what cross-activation looks like: those brain regions activate each other because the signals within their circuitry have a similar mathematical shape, independently of whether those signals represent an image, a sound or a movement. At the neural level they are, ultimately, just electrical waves.

This whole business of cross-activation is reminiscent of how mirror neurons link concepts across brain maps.

Ramachandran suspects that the inferior parietal lobe evolved originally for cross-modal abstraction, since it has to mediate signals coming from the touch, vision and hearing regions of the brain; and then later this feature became an independent skill, the ability to think abstract thoughts. The inferior parietal lobe (IPL) in the right hemisphere became skilled at bodily metaphors and the IPL in the left hemisphere become good at linguistic metaphors. The original abstraction was probably the mapping of vision into gesture, and to this day half of the IPL, the supramarginal gyrus, is responsible for coordinating vision and gesture (a fundamental feature of the human brain, that allows us to build and use tools to an extent unknown in other species). The other half of the IPL, the angular gyrus, found by accident (by exaptation) that this same process is useful to find similarities among different domains, e.g. for metaphorical thinking.

Initially, the human brain may have been more interested in making and using tools, and therefore evolved the skills to relate vision and gesture. By exaptation, this brain function ended up also being "translated" into the domain of communication. If this is indeed the case, then the whole process of building tools out of parts could be the precursor of and the indirect cause for the tree structure of linguistic syntax: sentences too are constructed out of parts. The original language may have been the language of building tools, not the language of speaking words.

The Evolution of Technology

The US journalist Kevin Kelly points out that there is another evolution at work. Kelly draws drawing a parallel between organic life and the life of technology. More precisely, he compares the biosphere with the "technium", the set of all interconnected technologies collectively created by humans. He argues that the evolution of the technium is driven by forces that are similar to the ones that drive the evolution of life. He disagrees with Gould: biological evolution does have a direction. The direction is towards complexity. Kelly believes the same kind of law and direction is at work in the technium.

Technology is inevitable. Many inventions were "invented" at the same time by different people in different places: those people were just the practical vehicle that technology used to emerge; but it would have emerged anyway. Basically, technology parasites on human minds in order to survive, reproduce and evolve, just like memes do.

The supremacy of technology is becoming obvious: we depend on technology. Our extended phenotype (to use another biological concept) contains an increasing component of technology without which we would not survive.

However, the net effect of the growing complexity and diversity of technology is, in Kelly's opinion, positive. What technology "wants" is progress towards more and more freedom for us. The more technology we have, the more choices we have, and the more freedom we enjoy. Technology increases our free will. Technology, in fact, allows human genius to emerge and prosper. Without technology the human race would not have had Mozart and VanGogh.

Kelly even develops a sort of ethics for technology. A technology can never be bad just like an animal species cannot be bad: they are the product of evolution. It makes no sense to ask whether a species is good or bad. It makes no sense to ask whether a technology is good or bad. Kelly thinks that we have a duty to maximize technologies in society at large, i.e. to invent everything that can be invented, because technology gives us more freedom.

Semiotics: Signs and Messages

Semiotics provides a different perspective to study the nature and origin of language.

Semiotics, founded in the 1940s by the Danish linguist Louis Hjelmslev, had two important precursors in the US philosophers Charles Peirce (whose writings were rediscovered only in the 1930s) and Charles Morris (who in 1938 had formalized a theory of signs).

Peirce reduced all human knowledge to the idea of "sign" and identified three different kinds of signs: the index (a sign which bears a causal relation with its referent); the icon (which bears a relation of similarity with its referent); and the symbol (whose relation with its referent is purely conventional). For example, the flag of a sporting team is a symbol, while a photograph of the team is an icon. Movies often make use of indexes: ashes burning in an ashtray mean that someone was recently in the room, and clouds looming on the horizon mean it is about to rain. Most of the words that we use are symbols, because they are conventional signs referring to objects.

Morris defined the disciplines that study language according to the roles played by signs. Syntax studies the relation between signs and signs (as in "the" is an article, "meaning" is a noun, "of" is a preposition, etc.). Semantics studies the relation between signs and objects ("Piero is a writer" means that somebody whose name is "Piero" writes books). Finally, Pragmatics studies the relation between signs, objects and users (the sentence "Piero is a writer" may have been uttered to correct somebody who said that Piero is a carpenter).

The Swiss linguist Ferdinand de Saussure introduced the dualism of "signifier" (the word actually uttered) and the "signified" (the mental concept). ("Semiology" usually refers to the Saussure-an tradition, whereas "semiotics" refers to the Peirce-an tradition. Semiotics, as opposed to Semiology, is the study of all signs).

The Argentine semiotician Luis Prieto studied signs, in particular, as means of communication. For example, the Braille alphabet and traffic signs are signs used to communicate. A "code" is a set of symbols (the "alphabet") and a set of rules (the "grammar"). The code relates a system of expressions to a set of contents. A "message" is a set of symbols of the alphabet that has been ordered according to the rules of the grammar. This is a powerful generalization: language turns out to be only a particular case of communication. A sentence can be reduced to a process of encoding (by the speaker) and decoding (by the listener).

The Hungarian linguist Thomas Sebeok views semiotics as a branch of communication theory that studies messages, whether emitted by objects (such as machines) or animals or humans. In agreement with René Thom, Sebeok thinks that human sign behavior has nothing special that can distinguish it from animal sign behavior or even from the behavior of inanimate matter. Sebeok also bridged genetics and linguistics, aware that the code of the grammar generates a language the way the genetic code generates a living body.

The US linguist Merlin Donald speculated on how the human mind developed. He argued that at the beginning there was only episodic thinking: early hominids could only remember and think about episodes. Later, they learned how to communicate and then they learned how to build narratives. Symbolic thinking came last. Based on this scenario, the Danish semiotician Jesper Hoffmeyer drew his own conclusions: in the beginning there were stories, and then, little by little, individual words rose out of them. Which implies that language is fundamentally narrative in nature; that language is corporeal, has to do with motor-based behavior; and that the unit of communication among animals is the whole message, not the word.

Hoffmeyer introduced the concept of "semiosphere", the semiotic equivalent of the atmosphere and the biosphere that incorporates all forms of communication, from smells to waves: all signs of life. Every living organism must adapt to its semiosphere or die. At all levels, life must be viewed as a network of "sign processes". The very reason for evolution is death: since organisms cannot survive in the physical sense they must survive in the semiotic sense, i.e. by making copies of themselves. "Heredity is semiotic survival".

Rene' Thom, the French mathematician who invented catastrophe theory, aimed to "geometrize thought and language". Thom was envisioning a Physics of meaning, of significant form, which he called "Semiophysics".

Following in this generalization of signs, James Fetzer ("Signs and Minds", 1988) even argued in favor of extending Newell and Simon's theory to signs: the mind not as a processor of symbols, but as a processor of signs.

Collective Cognition

What is, ultimately, the function of language? To communicate? To think? To remember? All of this and more. But, most likely, not only for the sake of the individual. Language's crucial function is to create a unit out of so many individuals. Once we learn to speak, we become part of something bigger than our selves. We inherit other people's memories (including the memories of people who have long been dead) and become capable of sharing our own memories with other people (even those who have not been born yet).

Thanks to language, the entire human race becomes one cognitive unit, with the ability to perceive, learn, remember, reason, and so forth. Language turns the minds of millions of individuals into gears at the service of one gigantic mind.

As the US neuroscientist Paul Churchland once pointed out, language creates a collective cognition, a collective memory and intelligence.

Further Reading

Bickerton, Derek & Calvin, William: LINGUA EX MACHINA (MIT Press, 2000)

Bickerton, Derek: LANGUAGE AND SPECIES (Chicago Univ Press, 1992)

Churchland, Paul: ENGINE OF REASON (MIT Press, 1995)

Darwin, Charles: LANGUAGES AND SPECIES (1874)

Deacon, Terrence: THE SYMBOLIC SPECIES (W.W. Norton & C., 1997)

Dennett, Daniel: KINDS OF MINDS (Basic, 1998)

Donald, Merlin: ORIGINS OF THE MODERN MIND (Harvard Univ Press, 1991)

Dunbar, Robin: GROOMING, GOSSIP AND THE EVOLUTION OF LANGUAGE (Faber and Faber, 1996)

Fetzer, James: COMPUTERS AND COGNITION (Kluwer Academic, 2001)
Gregory, Richard: MIND IN SCIENCE (Cambridge Univ Press, 1981)
Hoffmeyer, Jesper: SIGNS OF MEANING IN THE UNIVERSE (Indiana Univ. Press, 1996)
Hjelmslev, Louis: PROLEGOMENA TO A THEORY OF LANGUAGE (1943)
Humphrey, Nicholas: CONSCIOUSNESS REGAINED (Oxford Univ Press, 1983)
Jackendoff, Ray: PATTERNS IN THE MIND (Basic Books, 1994)
Jespersen, Otto: LANGUAGE (1922)
Joseph, Rhawn: NAKED NEURON (Plenum, 1993)
Kurth, Ernst: MUSIKPSYCHOLOGIE (1931)
Langer, Susanne: PHILOSOPHY IN A NEW KEY (1942)
Lieberman, Philip: THE BIOLOGY AND EVOLUTION OF LANGUAGE (Harvard Univ Press, 1984)
Miller, Geoffrey: THE MATING MIND (Doubleday, 2000)
Morris, Charles: FOUNDATIONS OF THE THEORY OF SIGNS (University Of Chicago Press, 1938)
Niehoff, Debra: THE LANGUAGE OF LIFE: HOW CELLS COMMUNICATE IN HEALTH AND DISEASE (Joseph Henry Press, 2005)
Oakley, Kenneth: MAN THE TOOL-MAKER (1949)
Ogden, Charles & Richards, Ivor: The Meaning of Meaning (1923)
Peirce, Charles: COLLECTED PAPERS (Harvard Univ Press, 1931)
Prieto, Luis: PRINCIPES DE NOOLOGIE (1964)
Saussure, Ferdinand: COURSE IN GENERAL LINGUISTICS (1916)
Sebeok, Thomas: APPROACHES TO SEMIOTICS (1964)
Taylor, Timothy: THE ARTIFICIAL APE (MacMillan, 2010)
Thom, René: SEMIOPHYSICS (Addison-Wesley, 1990)
Wilson, Frank: THE HAND (Pantheon Books, 1998)

Metaphor: How We Speak

Standing for Something Else

Metaphor is not just a poet's tool to express touching feelings. Metaphor is pervasive in our language. "Her marriage is a nightmare", "My room is a jungle", "She is a snake", "This job is a piece of cake", etc.: we communicate all the time, metaphorically.

The reason metaphor is so convenient is that it allows us to express a lot starting with very little: metaphor is a linguistic device to transfer properties from one concept to another.

Metaphor is so pervasive that every single word of our language may have originated from a metaphor. All language may be metaphorical. Given the importance of language among our mental faculties, metaphor is likely a key element of reasoning and thinking in general. In other words, being able to construct and understand metaphors (to transfer properties from a "source" to a "destination", from "nightmare" to "marriage", from "jungle" to "room", etc.) may be an essential part of being a mind.

Founding his theory on archeology, i.e. on evidence from prehistory, the British archeologist Steven Mithen became convinced that metaphor was pivotal for the development of the human mind.

A special case of metaphor is metonymy, which occurs when a term is used to indicate something else, e.g. "the White House" to mean "The President of the United States" rather than the building itself (as in "the White House pledged not to increase taxes"). Metonymy differs from metaphor in that metaphor is a way to conceive something in terms of another thing, whereas metonymy is a way to use something to stand for something else (i.e., it also has a "referential" function).

The study of metaphor presents a number of obvious problems: how to determine its truth value (taken literally, metaphors are almost always false) and how to recognize an expression as a metaphor (metaphors have no consistent syntactic form).

It is also intriguing that metaphor seems to violate so blatantly Paul Grice's conversational rules: if the speaker tries to make communication as "rational" as possible, why would she construct a metaphor instead of just being literal? The answer lies in the true nature of metaphor.

The Dynamics of Language

Early studies on metaphor focused on the analogical reasoning that metaphor implies.

The French philologist Michel Breal had already pointed out at the end of the 19th century, metaphor is often indispensable to express a concept for which words just do not exist in the language. Entire domains are mapped in other domains for lack of appropriate words. For example, the domain of character is mapped into the domain of temperature: a hot temper, a cold behavior, a warm person, etc. Breal realized that, eventually, metaphors shape language.

The British mathematician Max Black was influential in moving metaphor from the level of words to the level of concepts. His "interactionist" theory of metaphor ("Metaphor", 1955), inspired by the pioneering work of the British literary critic Ivor Richards in the 1930s, views metaphor not as a game of words, but as a cognitive phenomenon that involves concepts. In literal language, two concepts can be combined to obtain another concept without changing the original concepts (e.g., "good" and "marriage" form "good marriage"). In metaphorical language, two concepts are combined so that they form a new concept (e.g., marriage as a nightmare), and additionally they change each other (both "marriage" and "nightmare" acquire a different meaning, one reflecting the nightmarish aspects of marriage and the other one reflecting the marriage-like quality of a nightmare). They trade meaning. Predications that are normally applied to one are now also possible on the other, and vice versa. A metaphor consists in a transaction between two concepts. The interpretation of both concepts is altered.

Black viewed metaphor as a means to reorganize the properties of the destination. First of all, a metaphor is not an isolated term, but a sentence. A metaphorical sentence (e.g., "marriage is a nightmare") involves two subjects. The secondary subject (e.g., "nightmare") comes with a system of associated stereotyped information (or "predication"). That stereotyped information is used as a filter on the principal subject (e.g., "marriage"). There arises a "tension" between the two subjects of the metaphor. That tension is also reflected back to the secondary subject.

Black emphasized that metaphorizing is related to categorizing (the choice of a category in which to place an object is a choice of perspective), but is distinguished from it by an incongruity which causes a reordering and a new perspective.

A crucial point is that metaphor does not express similarities: it creates similarity.

Metaphors act on the organization of the lexicon and the model of the world.

Finally, Black argued that language is dynamic: over time, what is literal may become metaphoric and viceversa.

The Australian mathematician Michael Arbib, one of the many who have argued that all language is metaphorical, based his theory of language on Black's interactionist model.

At the other extreme, the US computer scientist James Martin does not believe that the process of comprehending a metaphor is a process of reasoning by analogy. A metaphor is simply a linguistic convention within a linguistic community, an "abbreviation" for a concept that would otherwise require too many words. There is no need for transfer of properties from one concept to another. In his theory, a number of primitive classes of metaphors (metaphors that are part of the knowledge of language) are used to build all the others. A metaphor is therefore built and comprehended just like any other lexical entity. Martin's is a purely "syntactic" model of metaphor.

Metaphorical Thought

The US linguist George Lakoff carried out a cognitive analysis of metaphor during the 1970s and 1980s. He reached two fundamental conclusions: (1) All language is metaphorical and (2) All metaphors are ultimately based on our bodily experience.

Metaphor shapes our language as well as our thought, and it does so by grounding concepts in our body. It provides an experiential framework in which we can accommodate abstract concepts. Thanks to metaphor, we can reduce (and therefore understand) abstract concepts to our physical experiences and to our relationship with the external world. Metaphor is therefore an intermediary between our conceptual representation of the world and our sensory experience of the world (an approach reminiscent of Immanuel Kant's schema).

Metaphor is not only ubiquitous in our language, it is also organized in conceptual systems: concepts of love are related to concepts of voyage, concepts of argument are related to concepts of war, and so forth. It is not only one word that relates to another word: it is an entire conceptual system that is related to another conceptual system.

This organization of conceptual systems forms a "cognitive map". Metaphor projects the cognitive map of a domain (the "vehicle") onto another domain (the "tenor") for the purpose of grounding the latter to sensory experience via the cognitive map of the former.

The entire conceptual castle of our mind relies on this creation of abstractions by metaphor from the foundations of our bodily experience in the world.

Lakoff grew up at a time when there was solid agreement about what metaphors are. Metaphor was merely considered a linguistic expression favored by poets that is not used in the literal sense and expresses a similarity. But he quickly started realizing that we use metaphors all the time, and that we use them in a far more encompassing manner. For example, we express love in terms of a journey (as in "our marriage isn't going anywhere"), or time in terms of money (as in "a waste of time"). Love is not similar to a journey, and time is not similar to money. Furthermore, abstract concepts (such as "love") are defined by metaphors. If we take the metaphors away all that is left are the roles (e.g., the lovers and the type of relationship). The system of metaphors built around an abstraction (e.g., all the metaphors that we use about love) tells us how to reason about that abstraction.

This led Lakoff to reason that: metaphor is not in the words, it is in the ideas; it is part of ordinary language, not only of poetry; it is used for reasoning.

Once metaphor is defined as the process of experiencing something in terms of something else, metaphor turns out to be pervasive, and not only in language but also in action and thought. The human conceptual system is fundamentally metaphorical in nature, as most concepts are understood in terms of other concepts. Language comprehension always consists in comprehending something in terms of something else. All our concepts are of metaphorical nature and are based on our physical experience.

Lakoff analyzed numerous domains of human knowledge and invariably detected the underlying metaphors. Theories, for example, are treated as buildings (a theory has "foundations" and is "supported" by data, theories are "fragile" or "solid"). Mathematics itself is a metaphor (trigonometry is a metaphor for talking about angles)

We understand the world through metaphors, and we do so without any effort, automatically and unconsciously. It doesn't require us to think, it just happens and it happens all the time. Most of the time we are thinking metaphorically without even knowing it.

Our mind shares with the other minds a conventional system of metaphor. This is a system of "mappings", of referring one domain of experience to another domain, so that one domain can be understood through another domain which is somehow more basic. Normally, a more abstract domain is explained in terms of a more concrete domain. The more concrete the domain, the more "natural" it is for our minds to operate in it.

Metaphors are used to partially structure daily concepts. They are not random, but rather form a coherent system that allows humans to conceptualize their experience. Again, metaphors create similarities.

Lakoff defined three types of metaphors: "orientational" (in which we use our experience with spatial orientation), "ontological" (in which we use our experience with physical objects), "structural" (in which natural types are used to define other concepts). Every metaphor can be reduced to a more primitive metaphor.

Language was probably created to deal only with physical objects, and later extended to non-physical objects by means of metaphors. Conceptual metaphors transport properties from structures of the physical world to non-physical structures.

Reason, in general, is not disembodied; it is shaped by the body.

Our conceptual system is shaped by positive feedback from the environment. As Edward Sapir and Benjamin Whorf had already argued before him, language reflects the conceptual system of the speaker.

Lakoff emphasized that metaphor is not only a matter of words, but also a matter of thought, that metaphor is central to our understanding of the world and the self.

Lakoff showed how a small number of metaphors could define a whole system of thought.

Even ritual was viewed by Lakoff as a crucial process in preserving and propagating cultural metaphors.

The reason that metaphor is so pervasive is that it is biological: our brains are built for metaphorical thought. Our brains evolved with "high-level" cortical areas taking input from "lower level" perceptual and motor areas. As a consequence, spatial and motor concepts are the natural basis for abstract reason. It turns out that "metaphor" refers to a physiological mechanism, to the ability of our brain to employ perceptual and motor inferential processes in creating abstract inferential processes. Metaphorical language is nothing but one aspect of our metaphorical brain.

Blending Metaphors

Joe Grady ("Foundations of meaning: primary metaphors and primary scenes", 1997) showed how complex metaphors are made of atomic metaphorical parts (or "primary metaphors") and these are, in turn, the product of cross-domain associations both at the individual and at the social level (that typically occur during the early stages of life). Atomic metaphorical parts are then "blended" in complex metaphors (as in Gilles Fauconnier's "conventional blending"). A metaphor results in the simultaneous activation of the constituent parts.

Thus we acquire metaphors all the time automatically and unconsciously during our daily life (children often do not separate two elements of an experience that always occur together until later in life). We then employ these metaphors in our daily life.

Primary metaphors include "affection is warm" (as in "a warm smile"), "important is big" ("a big opportunity"), "more is up" (high prices), "time is motion" ("time flies").

Metaphors are Fuzzy

The US philosopher Earl MacCormac advanced a unified theory of metaphor with broad implications for meaning and truth.

MacCormac rejected both the "tension" theory (which locates the difference between metaphor and analogy in the emotional tension generated by the juxtaposition of anomalous referents) and "controversion" theory pioneered by the US philosopher Monroe Beardsley (which locates that difference in the falsity produced by a literal reading of the identification of the two referents) and the "deviance" theory (which locates that difference in the ungrammaticality of the juxtaposition of two referents). MacCormac thinks that a metaphor is recognized as a metaphor on the basis of the semantic anomaly produced by the juxtaposition of referents. And this also means that metaphor must be distinct from ordinary language (as opposed to the view that all language is metaphorical).

MacCormac was influenced by the US philosopher Philip Wheelwright, who had classified metaphors into "epiphors" (metaphors that express the existence of something) and "diaphors" (metaphors that imply the possibility of something). Diaphor and epiphor measure the likeness and the dissimilarity of the attributes of the referents. A diaphor can become an epiphor (when the object is found to really exist) and an epiphor can become a literal expression (when the term has been used for so long that people have forgotten its origin).

Metaphor is a process that exists at three levels: a language process (from ordinary language to diaphor to epiphor back to ordinary language); a semantic and syntactic process (its linguistic explanation); and a cognitive process (to acquire new knowledge). Therefore a theory of metaphor requires three levels: a surface (or literal) level, a semantic level and a cognitive level.

The semantics of metaphor can then be formalized using the mathematical tool of fuzzy logic. Literal truth, figurality and falsity can be viewed as forming a continuum of possibilities rather than a discrete set of possibilities. The figurality of the metaphorical language, in particular, can be viewed as a continuum of "partial" truths that extends from absolute falsehood to absolute truth. These partial truths can be represented by fuzzy values.

This is expressed by a real number on a scale from zero to one: zero is absolute falsehood; the interval from zero to a certain value represents falsehood; the interval from that value to another value represents diaphor; the interval from that value to another value represents epiphor; and the last interval to one represents truth (with one representing absolute truth). Metaphors are born as diaphors and, as they become more and more familiar through commonplace use, slowly mutate into epiphors, thereby losing their "emotive tension".

Language can then be represented mathematically as a hierarchical network in n-dimensional space with each of the nodes of the network a fuzzy set (defining a semantic marker). When unlikely markers are juxtaposed, the degrees of membership of one semantic marker in the fuzzy set representing the other semantic marker can be expressed in a four-valued logic (so that a metaphor is not only true or false).

MacCormac argued that, as cognitive processes, metaphors mediate between culture and mind, influencing both cultural and biological evolution.

Metaphor as Meaning

Drawing from Black's interactionist theory, and its vision of metaphor's dual content (literal and metaphorical, "vehicle" and "topic"), as well as from Ferdinand de Saussure's theory of signs, the US philosopher Eva Kittay developed a "relational" theory of meaning for metaphor. Her principle was that the meaning of a word is determined by other words that are related to it by the lexicon. Meaning is not an item: it is a field. A semantic field is a group of words that are semantically related to each other. Metaphor is a process that transfers semantic structures between two semantic fields: some structures of the first field create or reorganize a structure in the second field.

The meaning of a word consists of all the literal senses of that word. A literal sense consists of a conceptual content, a set of conditions, or semantic combination rules (permissible semantic combinations of the word, analogous to Fodor's selection-restriction rules), and a semantic field indicator (relation of the conceptual content to other concepts in a content domain). An interpretation of an utterance is any of the senses of that utterance. Projection rules combine lower-level units into higher-level units according to their semantic combination rules. A first-order interpretation of an utterance is derived from a valid combination of the first-order meanings of its constituents. Second-order interpretation is a function of first-order interpretation and expresses the intuitive fact that what has to be communicated is not what is indicated by the utterance's literal meaning.

Semantic fields help to recognize an utterance as a metaphor. For example, an explicit cue to the metaphorical nature of an utterance is when the first-order and the second-order interpretation point to two distinct semantic fields.

The cognitive force of a metaphor comes from a re-conceptualization of information about the world that has already been acquired but possibly not conceptualized.

Ultimately, Kittay thinks that metaphorical meaning is not reducible to literal meaning. Metaphor is, de facto, second-order meaning.

Metaphor As Physiological Organization

Borrowing from the work of the Hungarian linguist Stevan Harnad, who thinks that sensory experience is recorded as a continuous engram whereas a concept derived from sensory experience is recorded by discrete engrams ("Metaphor and Mental Duality", 1982), the Indian mathematician Bipin Indurkhya thinks that metaphor originated from the interaction between sensory-based and concept-based representations in the brain, which are structurally different.

He focuses on how metaphor creates similarity. This is, in itself, a paradox. By definition, a metaphor implies that there are at least two different ways to represent a situation. At the same time, we assume that these different representations are not arbitrary but they somehow "interact" (are coherent). Indurkhya points out three level of reality. The "God's eye view" of the world is independent of any cognitive being perceiving it. It's the world as it is. Cognitive beings interact with it via their sensorimotor system and the interaction yields the second level of reality, whose "ontology" depends on the sensorimotor system of the agent. Since cognitive agents have different sensorimotor systems , no surprise that different cognitive agents perceive a different reality.

The third level is the network of concepts created and used by the cognitive agent. A cognitive model is the mapping of this network into sense-data. The same mind can create different representations of the same reality, yielding different cognitive models. As the cognitive being "grows", there are two ways that it can maintaining coherence. It can restructure the network of concepts to better accommodate new data, a process that frequently creates new concepts; or it can change the mappings from the network of concepts to the sense-data, and basically "reuse" the existing concepts. The latter process is called "projection". So the cognitive agent "grows" via a dual process of accommodation and projection.

The latter process is the one that engenders metaphors. A metaphor is the projection of one conceptual network (the source) into another conceptual network (the target). Some concepts of the source maintain their conventional interpretation (the way the cognitive system usually interprets them) but others will require an unconventional ("metaphorical") interpretation.

Metaphor As Conceptual Blending

Gilles Fauconnier's theory of mental spaces constitutes a natural generalization of metaphor, in which the number of conceptual spaces that blend is only two: the one described by the metaphor (the "tenor") and the one which provides the description (the "vehicle").

Fauconnier adds two more spaces to deal with metaphors: the "generic" space, which represents all the shared concepts that are required by tenor and vehicle, which are necessary not only to understand the metaphor but to mediate between the tenor and the vehicle; and the "blend" space, which contains the solution, the concept generated by the metaphor. The tenor and the vehicle are reconciled thanks to the generic space and this reconciliation produces the blend space.

Lying Helps Communication

We cheat on children. Every time we tell children a fairy tale, we are lying to them. Those characters do not exist. Santa Claus does not exist. The cartoons on tv do not exist. We tell them lies all the time. And, still, children "learn". What they learn is not the literal meaning of those stories. In fact, they themselves frequently doubt those stories. But children understand that what they are supposed to learn is not the literal meaning. Children somehow understand that their parents are trying to teach them something else. Children somehow understand that parents use fairy tales because it is a more efficient and painless way to teach them what really matters. What matters is not that Little Red Riding Hood met a wolf, but that there are good and bad people. Children's brains are programmed to somehow discard the fairy tale and grasp the "meaning" of the story.

Long after they forget what Little Red Riding Hood was doing in the woods they will still remember that there are good and bad people.

As we grow up, we assume that we stop lying to each other. As adults, we try to tell it as it is. That might be true for scientists, but it is hardly visible in ordinary lives. Every argument between two people usually involves some kind of exaggeration. Most political discussions start with unreasonable statements (a popular one like "depleted uranium killed one million Iraqi children" is obviously false if one checks the population of Iraq but it was widespread all over the world in the 1990s). And most people routinely exaggerate the details that most matter within a story. The listener, on the other hand, routinely "decodes" those exaggerations. Language lends itself to a level of ambiguity that we use to deliver more than the literal meaning. Thus adults tell each other "fairy tales". It never truly ends. Even scientists, to some extent, exaggerate the implications of their theories when they try to explain them to ordinary people. A distortion of reality seems to be useful, if not essential, to human communication.

Metaphorical Ignorance
There is one weakness in the experimental praxis of linguists: they only study people who are fluent in a language. If you want to study Chaxipean, you go to Chaxipe and talk to Chaxipeans. They are the world experts in Chaxipean. Most of our ideas on language, categorization and metaphors come from studying people who are fluent in a language.

But the brain of a person who is not fluent in that language should be working the same way. My "use" of the German language, though, is not the same as a native German's. I stay away from metaphors in a language like German that i barely understand. Using a metaphor in German sounds scary to me. If i am speaking in a foreign language, i stick to simple sentences whose meaning is transparent. I do not say "their marriage is going nowhere": i say "their marriage is not good". I reduce all concepts to elementary concepts of good and bad, ugly and beautiful, etc. My mastery of the foreign language is not such that I can afford to use metaphorical expressions.

This goes against the claim that metaphor is useful to express meaning in a more efficient way. People who do not master a language should use metaphor precisely to compensate that deficiency. Instead, we tend to do the opposite: if we do not master a language, we avoid metaphors. Metaphorical language requires mastering the language skills first, and is proportionate to those skills. This is what the traditional theory predicted (metaphor is for poets, language specialists). There was a grain of truth in it.

Further Reading
Arbib, Michael: CONSTRUCTION OF REALITY (Cambridge University Press, 1986)

Black, Max: MODELS AND METAPHORS (Cornell Univ Press, 1962)

Fauconnier, Gilles: MENTAL SPACES (MIT Press, 1994)

Breal, Michel: ESSAI DE SEMANTIQUE (1897)

Hintikka, Jaakko: ASPECTS OF METAPHOR (Kluwer Academics, 1994)

Kittay, Eva: METAPHOR (Clarendon Press, 1987)

Indurkhya, Bipin: METAPHOR AND COGNITION (Kluwer Academic, 1992)

Lakoff, George: METAPHORS WE LIVE BY (Chicago Univ Press, 1980)

Lakoff, George: MORE THAN COOL REASON (University of Chicago Press, 1989)

Lakoff, George: WOMEN, FIRE AND DANGEROUS THINGS (Univ of Chicago Press, 1987)

Lakoff, George: PHILOSOPHY IN THE FLESH (Basic, 1998)

MacCormac, Earl: A COGNITIVE THEORY OF METAPHOR (MIT Press, 1985)

Martin, James: A COMPUTATIONAL MODEL OF METAPHOR INTERPRETATION (Academic Press, 1990)

Mithen Steven: THE PREHISTORY OF THE MIND (Thames and Hudson, 1996)

Ortony, Andrew: METAPHOR AND THOUGHT (Cambridge Univ Press, 1979)

Richards, Ivor: THE PHILOSOPHY OF RHETORIC (Oxford Univ. Press, 1936)

Wheelwright, Philip: METAPHOR AND REALITY (1962)

Pragmatics: What We Speak

The Use of Language

The US linguist Michael Reddy ("The Conduit Metaphor", 1979) dubbed "conduit metaphor" the idea, deeply entrenched in popular thinking, that the mind contains thoughts that can be treated like objects. That idea views linguistic expressions as vehicles for transporting ideas along a conduit which extends from the speaker to the listener. These vehicles are strings of words, each of which contains a finite amount of a substance called meaning: the speaker assembles the meaning, loads the vehicle and sends it along the conduit. The listener receives the vehicle, unloads it and unscrambles the meaning. This "conduit metaphor" is widespread in the languages we speak. Reddy thinks otherwise: the transfer of thought is not a deterministic, mechanical process. It is an interactive, cooperative process.

Language is a much more complicated affair than it appears. Syntax, metaphor, semantics are simply aspects of how we interpret and construct sentences. But first and foremost language is a game in which we engage other speakers. A lot more information is exchanged through the "use" we make of language.

In the 1940s the Austrian philosopher Ludwig Wittgenstein had argued that to understand a word is to understand a language and to understand a language is to master the linguistic skills. A word, or a sentence, has no meaning per se. It is not the meaning, it is the "use" of language that matters.

The discipline of "pragmatics" studies aspects of meaning that cannot be accounted for by semantics alone, but have to do with the way language is used. Ultimately, the pragmatic goal of language is to understand the "reason" of a speech. What are the speaker's motive and goal? For example, semantics can account for the meaning of the sentence "do you know what time it is?", but not for the fact that an answer is required (the speaker's intention is to learn what time it is).

The way language is used has to do with the context: there is no speech without a context. For example, the same sentence may be used for two different purposes in two different contexts: "do you know what time it is?" may be a request (equivalent to "what time is it?") or a reproach (as in "you are very late"). The only way to discriminate what that sentence really means is to analyze the context.

Ultimately, meaning arises from the relationship between language and context. Because context is the key element, pragmatic studies focus on "indexicals", "implicatures" and "presuppositions".

Indexicals are terms such as "i", "today", "now" whose referents depend on the context: "I am a writer" is true if i am Piero Scaruffi, but it is false if i am Elvis Presley. Just like "i" may refer to any person in the world, "today" may refer to any day of the year and "here" to any place in the universe. Only the context can fix the meaning of indexicals.

Implicatures are the facts that are implied by the sentence: "the Pope held mass in St. Peter's square" implies that the Pope is alive.

Presuppositions are the facts that are taken for granted, for example the fact that humans die and a job earns money.

The purpose of a speech in a given context is to generate some kind of action. There is an "intention" to the speech and to the way the speech is structured. Pragmatics studies intentional human action. Intentional action is action intended to achieve a goal, through some kind of plan, given some beliefs about the state of things. This intention results in "speech acts" that carry out, directly or indirectly, the plan. Therefore the pragmatic dimension of language deals with beliefs, goals, plans and ultimately with speech acts, unlike syntax and semantics which deal, respectively, with the structure of a sentence and with the isolated meaning of the sentence.

More broadly, the answer to a question is not simply the information requested by the interrogator but a move in a tactical game based on the responder's expectations about the interrogator's intentions. It is not knowledge that you deliver but a strategy of how to use your knowledge to affect the people around you. Ultimately, Pragmatics borders on Psychology.

Language as Cooperation

In 1967 the British philosopher Paul Grice had a key intuition: that language is based on a form of cooperation among the speakers. For language to be meaningful, both the speaker and the hearer must cooperate in the way they speak and in the way they listen. The way they do it, is actually very simple: people always choose the speech acts that achieve the goal with minimum cost and highest efficiency.

Language has meaning to the extent that some conventions hold within the linguistic community. Those conventions help the speaker achieve her goal. The participants of a conversation cooperate in saying only what makes sense in that circumstance.

Grice focused on the linguistic interplay between the speaker, who wants to be understood and cause an action, and the listener. This goes beyond syntax and semantics. A sentence has a timeless meaning, but also an occasional meaning: what the speaker meant to achieve when she uttered it.

Grice's four maxims summarize those conventions: provide as much information as needed in the context, but no more than needed (quantity), tell true information (quality), say only things that are relevant to the context (relation), avoid ambiguity as much as possible (manner).

The significance of an utterance includes both what is said (the explicit) and what is implicated (the implicit). Grice therefore distinguished between the "proposition expressed" from the "proposition implied", or the "implicature". Implicatures exhibit properties of cancellability (the implicature can be removed without creating a contradiction) and calculability (an implicature can always be derived by reasoning under the assumption that the speaker is observing pragmatic principles).

Grice's maxims help the speaker say more than what she is actually saying. They do so through implicatures that are implied by the utterance. Grice distinguishes two types of implicatures, depending on how they arise. Conventional implicatures are determined by linguistic constructions in the utterance. Conversational implicatures follow from maxims of "truthfulness", "informativeness", "relevance" and "clarity" that speakers are assumed to observe. Conversational implicatures can be discovered through an inferential process: the hearer can deduce that the speaker meant something besides what he said by the fact that what he said led the hearer to believe in something and the speaker did not do anything to stop him from believing it.

The fundamental intuition was that there is more to a sentence than its meaning. A sentence is "used" for a purpose.

The US linguist Jerrold Sadock even distinguishes semantic sense from interpreted sense, i.e. meaning from use, as two different aspects of language. Arguing that humans employ a number of different skills during their linguistic acts, Sadock believes that one needs not just one grammar but a set of autonomous pseudo-grammars, each devoted to one dimension of language.

The Logic of Speech Acts

In the 1950s the British Philosopher John-Langshaw Austin had started a whole new way of analyzing language by viewing it as a particular case of action: "speech action".

Austin introduced a tripartite classification of acts performed when a person speaks. Each utterance entails three different categories of speech acts. The "locutionary" act consists of the words employed to deliver the utterance. The "illocutionary" act is determined by the type of action that the utterance performs, such as warning, commanding, promising, asking. The "perlocutionary" act is the effect that the act has on the listener, such as believing or answering.

A locutionary act is the act of producing a meaningful linguistic sentence. An illocutionary act sheds light on why the speaker is uttering that meaningful linguistic sentence. A perlocutionary act is performed only if the speaker's strategy succeeds.

The "locution" is the act of saying something. That, in turn, can be dissected into three acts. The physical movement that causes sounds to be produced is the "phonetic" act, each specific phonetic act being a "phone". The fact that that utterance also conforms to the linguistic rules of a specific language is its "phatic" act, each specific phatic act being a "pheme". The fact that the pheme also referred to some people, objects and situations is a "rheme", a "rhetic" act. A rheme requires a pheme and a phone. Therefore, rhemes are a sub-class of phemes, which in turn are a sub-class of phones. A phonetic act fails if there is nobody listening, a phatic act fails if the listener does not understand the language of the speaker or if the speaker makes grammatical mistakes, and a rhetic act fails if the speaker does not adequately deliver the meaning he had in mind.

Illocutionary acts are performed by a speaker when she utters a sentence with certain intentions (e.g., statements, questions, commands, promises).

Austin believed that any locutionary act (phonetic act plus phatic act plus rhetic act) is part of a discourse which bestows an illocutionary force on it. All language is therefore an illocutionary act.

In the 1970s the US philosopher John Searle developed a formal theory of the conditions that preside over the genesis of speech acts. Searle classifies such acts in several categories, including "directive acts", "assertive acts", "permissive acts" and "prohibitive acts". And showed that only assertive acts can be treated with classical logic. An illocutionary act consists of an illocutionary force (e.g., statement, question, command, promise) and a propositional content (what it says). Illocutionary acts are the minimal units of human communication, and argued that the illocutionary force of sentences is what determines the semantics of language.

The Logic of Relevance

The French sociologist Dan Sperber and the British linguist Deirdre Wilson have shown that "relevance" constrains the coherence of a discourse and enables its understanding. Relevance is a relation between a proposition and a set of contextual assumptions: a proposition is relevant in a context if and only if it has at least one contextual implication in that context. The contextual implications of a proposition in a context are all the propositions that can be deduced from the union of the proposition with the context.

A universal goal in communication is that the hearer is out to acquire relevant information. Another universal goal is that the speaker tries to make his utterance as relevant as possible. Understanding an utterance then consists in finding an interpretation that is consistent with the principle of relevance. The principle of relevance holds that any act of ostensive communication also includes a guarantee of its own optimal relevance. This principle is proven to subsume Grice's maxims.

Relevance can arise in three ways: an interaction with assumptions that yields new assumptions; the contradiction of an assumption which removes it; additional evidence for an assumption that strengthens the confidence in it.

Implicatures are either contextual assumptions or contextual implications that the hearer must grasp to recognize the speaker as observing the principle of relevance. The process of comprehending an utterance is thus reduced to a process of hypothesis formation and confirmation: the best hypothesis about the speaker's intentions and expectations is the one that best satisfies the principle of relevance.

Language can make sense only if speaker and listener cooperate. The US philosopher Donald Davidson points out that language transmits information. The speaker and the listener share a fundamental principle to make such transmission as efficient as possible. Such "principle of charity" (originally introduced by the US philosopher Neil Wilson) asserts that the interpretation to be chosen is the one in which the speaker is saying the highest number of true statements. During the conversation the listener tries to build an interpretation in which each sentence of the speaker is coupled with a truth-equivalent sentence.

Language, far from being a mechanical process of constructing sentences and absorbing sentences, is a subtle process of cooperating with the "other" to achieve the goal of communicating.

Narrating

One of the things that we do with language is to tell stories. In fact, one wonders if we do anything else. Even the simplest of communication or discussion involves many mini-stories.

The US psychologist Jerome Bruner pointed out that there are, basically, two kinds of thinking: the paradigmatic and the narrative. And they are like two different substances in that they represent the world in two different ways and they obey two different sets of laws.

They are irreducible to one another. One is reasoning, and the other one is narrating. One produces logical arguments whose goal is truth. The other one produces stories whose goal is plausibility. Abstract form is the key element of the former, whereas human psychology is the key element of the latter.

Bruner points out that human civilization has developed sophisticated analyses of how to think in the paradigmatic way (for example, mathematical Logic), but has little to say about how to think in the narrative way (how to write good stories).

Bruner believes that narrative thinking incorporates two dimensions: the "landscape of action" (the plot) and the "landscape of consciousness" (the motivations). The former outlines the actions and the actors, the latter outlines their mental states (goals, beliefs, emotions).

Following Lev Vygotsky, Bruner thinks that reality belongs to two spheres, the natural and the social, the former being more aptly described by paradigmatic thought (the sciences) and the latter being more aptly described by narrative thought.

Further Reading

Austin, John Langshaw: HOW TO DO THINGS WITH WORDS (Oxford Univ Press, 1962)

Bruner, Jerome: ACTUAL MINDS POSSIBLE WORLDS (Cambridge Univ Press, 1986)

Davidson, Donald: INQUIRIES INTO TRUTH AND INTERPRETATION (Clarendon Press, 1974)

Davis, Steven: PRAGMATICS (Oxford University Press, 1991)

Gazdar, Gerald: PRAGMATICS (Academic Press, 1979)

Green, Georgia: PRAGMATICS (Lawrence Erlbaum, 1996)

Grice, Paul: STUDIES IN THE WAY OF WORDS (Harvard Univ Press, 1967 lectures published in 1989)

Levinson, Stephen: PRAGMATICS (Cambridge Univ Press, 1983)

Sacks, Harvey: "Lectures on Conversation, Volumes I and II" (Wiley, 1967)

Sadock, Jerrold: TOWARD A LINGUISTIC THEORY OF SPEECH ACTS (Academic Press, 1974)

Sadock, Jerrold: AUTOLEXICAL SYNTAX (Univ. of Chicago Press, 1991)

Searle, John: SPEECH ACTS (Cambridge Univ Press, 1969)

Searle, John: EXPRESSION AND MEANING (Cambridge Univ Press, 1979)

Sperber, Dan & Wilson, Deirdre: RELEVANCE, COMMUNICATION AND COGNITION (Blackwell, 1995)

Wittegenstein, Ludwig: PHILOSOPHICAL INVESTIGATIONS (1958)

MEANING: A JOURNEY TO THE CENTER OF THE MIND

The Meaning Of Meaning

Meaning is intuitively very important. We all assume that what matters is the meaning of something, not the something per se. But then nobody really knows how to define what "meaning" means. The symbol "LIFE" per se is not very interesting, but the thing it means is very interesting.

Before we can answer the question "what is the meaning of life?" we need to answer to the simpler question: "what is meaning?" What do we mean when we say that something means something else?

Linguistic principles are innate and universal. Everybody is born with the ability to learn language. But our language ability depends not only on symbolic utterance (on constructing grammatically-correct sentences), but also on our ability to make use of symbolic utterance, to connect the idea with some action. Spoken language is not even that essential: we can communicate with gestures and images. This happens since birth, as linguistic and nonlinguistic information (e.g., visual) are tied from the beginning. What matters is the ability to comprehend. As a matter of fact, in a baby comprehension is ahead of utterance.

Since meaning is first and foremost "about" something, an obvious component of meaning is what philosophers call "intentionality": the ability to refer to something else. The word "LIFE" refers to the phenomenon of life. Far from being an exclusive of the human mind, meaning in this broad sense is rather pervasive in nature. One could even claim that everything in nature refers to everything else, as it would probably not exist without the rest of the universe. Certainly, a shadow refers to the object that makes it, and a crater refers to the meteor that created it. But "meaning" in the human mind also involves being aware of it, and in this narrow sense "meaning" could be an exclusive of the human mind. There are, therefore, at least two components of meaning: intentionality and awareness. Something refers to something else, and I am aware of this being the case.

At the same time, meaning affects two complementary aspects of our mind: language (i.e., the way we communicate meaning to other thinking beings) and concepts (i.e., the way we store meaning for future reference). As the British philosopher Paul Grice noted, there are two different meanings to the word "meaning": "what did she mean" and "what does that refer to". The meaning of a speaker is what she intended to say. There is an intention to take into account. The meaning of a word relating to a natural phenomenon is what it refers to (the meaning of "LIFE" is the phenomenon of life).

There is another dimension to meaning. Think of water, a fairly innocent subject: what does the concept of "water" refer to? The substance? The chemical compound H2O? Something liquid and transparent that is found in rivers and in the rain? Imagine that in another world there is a substance that looks and behaves just like water, but is made of a different chemical compound. When I and a person of that world think of water, are we referring to the same thing or to two different things? What is a clock? An object whose function is that of marking the time (which could be a sundial)? Or an object whose structure is round, has two hands and 12 numbers (which could be a toy clock that does not perform any actual function)?

Intension And Extension

A fundamental step in the discussion of meaning was Gottlob Frege's distinction between "sense" and "reference", which led to the distinction between "intension" and "extension". The "referent" of a word is the object it refers to, the "sense" of that word is the way the referent is given. For example, "the star of the morning" and "the star of the evening" have two different senses but the same referent (they both refer to the planet Venus). A more important example: propositions of classical Logic can only have one of two referents, true or false.

The "extension" of a concept is all the things that belong to that concept. For example, the extension of "true" is the set of all the propositions that are true. The "intension" of that concept is the concept itself. For example, the extension of "red" is all the objects that are red, whereas the intension of "red" is the fact of being red. There is an intuitive relationship between sense and intension, and between reference and extension.

But the relationship between sense and reference is not intuitive at all, as proved by the difficulty in handling indexicals (words such as "i") and demonstratives (such as "this"). The proposition "I am Piero Scaruffi" is true or false depending on who utters it. The proposition "I am right and you are wrong" has two completely opposite meanings depending on who utters it.

A number of alternatives to Frege's analysis have been proposed over the decades: Saul Kripke's and Hilary Putnam's "causal theory of reference" (which assumes a causal link between a word and what it stands for); Kripke's distinction of "rigid designators" and "non-rigid designators" in the context of possible worlds; and Richard Montague's intensional-logic approach (in which the sense of an expression is supposed to determine its reference). These are all different views on how sense and reference relate.

Meaning Is What We Do With It
Ludwig Wittgenstein introduced a different way of thinking about meaning, a common-sense way that identifies meaning with use. The meaning of something is what we do with it. This led to Conceptual/functional/inferential/procedural role semantics.

Similarly, proof-theoretic semantics, spearheaded in the 1940s by the German mathematician Gerhard Gentzen, takes the meaning of a proposition to be defined by the role that the proposition plays within the system of inference.

Does Meaning Exist?
"Nominalism" is a centuries-old philosophical faith according to which "universals" exist only in our mind and language. We tend to assign them a reality of their own when, in reality, they are just conventions used by our minds and our language.

At the other end of the spectrum, Plato claimed that ideas exist in a world of their own, independent of our material world.

Truth
What is a theory of truth? Let's take an example from Physics, a science that is famous for theories. A theory of electricity is an explanation of the nature and cause of electricity and a set of laws that electrical phenomena obey. A theory of truth is essentially an explanation of the nature of truth and a set of laws that "true" things obey.

Electricity is the property that all electrical things share. What is the property that all true statements have in common?

Why is a theory of truth important? Because that is what, ultimately, our cognitive life is all about: truth. Whenever we analyze a scene, whenever we analyze a statement, whenever we recall a memory, whenever we do anything with our brain, we are on a quest for truth.

Our cognitive life is a continuous struggle for truth: is that stain in the distance a tree? Is she home tonight? Will my flight take off on time? Why did the Roman empire fall?

Our mind, ultimately, is an organ to identify truth. The meaing of our life is truth.

The most intuitive theory of truth is perhaps the "correspondence theory of truth", that relates truth to reality: a statement is true if and only if the world it describes is real. Truth corresponds to the facts. The statement "Snow is white" is true in virtue of the fact that snow is, indeed, white. The statement "My name is Piero" is true in virtue of the fact that my name is, indeed, Piero. And so forth.

There are several problems with this theory of truth. The truth predicate (the expression "is true") acts as an intermediary between words (language, mind) and the world. One problem is that this definition of truth relates two things that are very different in nature and it is not clear how we can find a correspondence between things that belong to different realms. Precisely, statements (such as "Snow is white" and "My name is Piero") are mental objects. They are in my head. The reality we compare them with is made of objects, such as snow. A statement is made of a number of words (each of which may present its own problems at close scrutiny). The reality it refers to is made of objects and properties of objects. Is there truly a correspondece between the words "Snow is white" and the fact that snow is white? How can we compare two things that are different in nature, such as a mental object and a physical fact?

This point is important because we are supposed to define truth outside us: truth must not depend on us, it must depend on the world. Something is true not because i think so, but because there is some objective truth out there in the world. If this is the case, the problem is: how can a mental object like a statement relate to an object that is outside the mind.

Second, most statements just do not accurately reflect reality: is snow white? Not really. The closer you look at snow the less white it is. Is today a "hot" day? Yes, if you don't start arguing about which temperature qualifies for hot.

The first objection can be answered by observing that, if you believe in modern science, we rarely talk about things that exist and mostly talk about things that our brain presents us with. I don't know if there exists snow. My brain shows me something that we named snow, but quantum physics tells me that there is only a clod of particles. I see white, but quantum physics tells me that there is a stream of photons. And so forth. When we say that snow is white, we are not referring to something that exists in the world (it may or it may not exist). We are referring to something that is happening in our brain: our brain received some inputs from the senses and generated the perception of snow and of white. Therefore, both statements and "reality" are mental objects, and it is perfectly legitimate to relate a mental object such as the statement "Snow is white" to a mental object such as the perception or the memory that snow is white.

We can rephrase the correspondence problem in neural terms. A statement (for example, about the snow being white) is a neural pattern in the brain. The fact that snow is white is also a neural pattern in the brain (either a pattern of recalling a memory of snow or a pattern of perceiving the snow). It is perfectly legitimate to compare two patterns of brain activity.

The Polish mathematician Alfred Tarski found his own solution to the problems of the correspondence theory.

Alfred Tarski's theory of truth has two components. First, he defines a true statement as a statement that corresponds to reality. This is only a definition of "true statement" and not of "truth" in general. Of course, if one lists all true statements, one gets a definition of truth: truth is "Snow is white" and "My name is Piero" and "The Earth is not the center of the universe" and "France won the 1998 world cup" and "..." But this is neither elegant nor practical (most languages have an infinite number of true statements).

The second component to Tarski's theory is the idea that truth can only be defined relative to another language. Most languages include the word "true", but that leads to paradoxes such as "I am lying" which is both true and false at the same time. The problem is simply that "true" is a word of the language and we are applying it to a statement of the language. Tarski realized that one could not define truth in a language through the language itself and avoid contradictions. So he introduced a "meta-language" to define truth in the "object language". Truth in the object language can then be defined recursively from the truth of elementary statements (the "sentential functions"). "For all sentences s in language L, s is true if and only if T(s) is true", where T(s) is a formula containing s and L's primitives.

Tarski based his "model-theoretic" semantics (models of the world yield interpretations of sentences in that world) on this "correspondence theory of truth" (a statement is true if it corresponds to reality), so that the meaning of a proposition is, basically, the set of situations in which it is true.

Alas, Tarski's theory of truth does not work well with ordinary languages, although it works wonders with the formal languages of Logic.

The problem with Tarski's theory is that it is not clear what he defined. He did not define truth, but "truth in a language". By this, it is not clear if he indirectly acknowledged that the nature of truth is impossible or even pointless.

A secondary problem is that his theory does not distinguish the linguistic theory from the metaphysical theory: explaining the word "true" is a linguistic matter, whereas explaining the nature of truth is a metaphysical matter.

Tarski's theory is really about the linguistic feature.

Possible World Semantics

The US philosopher Saul Kripke expanded Tarski's model-theoretic interpretation to Modal Logic ("Semantical Analysis of Modal Logic",

1963). Modal Logic is a logic that adds two more truth values, "possible" and "necessary" (also known as "modal" values) to the two traditional ones, "true" and "false".

Kripke defined modality through the notion of possible worlds: a property is necessary if it is true in all worlds, a property is possible if it is true in at least one world. Thanks to these two operators, it is possible to discriminate between sentences that are false but have different intension. In classical Logic, sentences such as "Piero Scaruffi is the author of the Divine Comedy" and "Piero Scaruffi is a billionaire" have the same extension, because they are both false. In Modal Logic they have different extensions, because the former is impossible (because I was not alive at the time), whereas the latter is also false but could be true. Also, Modal Logic avoids paradoxes that classical Logic cannot deal with. For example, the sentence "all mermaids are male" is intuitively false, but classical Logic would consider it true (because the sentence "all mermaids are male" translates into a logical formula of the negation of something that is not true, i.e. that is always true). In Modal Logic this sentence is false in the world where mermaids do exist.

The advantage of Kripke's semantics is that it can interpret sentences that are not extensional (that do not satisfy Leibniz's law), such as those that employ opaque contexts (to know, to believe, to think) and those that employ modal operators. Put bluntly, Kripke's semantics can interpret all sentences that can be reduced to "it is possible that" and "it is necessary that". The trick is that in his semantics a statement that is false in this universe can be true in another universe. The truth values of a sentence are always relative to a particular world. A proposition does not have a truth value, but a set of truth values, one for each possible world.

Tarski's theory is purely extensional (for each model the truth of a predicate is determined by the list of objects for which it is true), whereas Kripke's modal logic is intensional. An extensional definition would actually be impossible, since the set of objects is infinite.

Kripke's semantics can explain how we can refer to a thing by its name, even when we do not know the properties of that thing.

John Stuart Mill introduced the concepts of "connotation" and "denotation". The expression "the president of Russia" denotes Vladimir Putin but it can be understood even by those who don't know who the current president of Russia is (i.e., it has a meaning regardless of whether you know that the president of Russia is Putin or not). Mill then divided names into general and singular (the latter including proper names). He then argued that general terms denote and connote, whereas singular terms merely denote. The word "Piero" simply refers to me, but it means nothing. Proper names have no meaning, general names have connotation (meaning). Frege and Russell interpreted proper names differently, as abbreviations of descriptions, and then the description is the sense (meaning) of the name ("sense" being Frege's term for connotation, and "reference" being Frege's term for denotation). In this view all names have connotation/sense. Kripke objected that Frege's "sense" is both the meaning of a name and the way its reference is determined. Kripke distinguishes the meaning of a designator and the way its reference is determined (which are both "sense" in Frege). In Kripke's view, names are rigid designators (their meaning is the same in every possible world); definite descriptions such as "the president of Russia", on the other hand, may have a different meaning in different worlds.

Kripke replaces Frege's description with a casual chain of communication Initially, the reference of a name is fixed by some operation (e.g., by description or by baptism), then the name is passed from person to person, from generation to generation. A name is not identified by a set of unique properties satisfied by the referent (Frege's description): the speaker may have erroneous beliefs about those properties or they may not be unique. A property cannot determine the reference as the object might not have that property in all worlds. Instead, the name is passed to the speaker by tradition from link to link. And names for natural kinds behave in a similar way to proper names. Natural kinds and proper names are not as different as traditionally assumed. Names are linked to their referents through a casual chain. A term applies directly to an object via a connection that was set in place by the initial naming of the object. A speaker is always a member of a community of speakers who use that name. New discoveries do not change the meaning of names: if we discovered that we have always been wrong about the chemical composition of gold, the name "gold" would still mean gold. If it turned out that water is not H_2O, i would still recognize water as water. The term "water" still designates water in a world in which water is not made of H_2O.

Neither proper nor common nouns are associated with properties that serve to select their referents. Names are just "rigid designators". Both proper and common names have a referent, but not a Fregean sense.

Analogously, Jerry Fodor argued in favor of two types of meaning: one is the "narrow content" of a mental representation, which is a semantic representation and is purely mental and does not depend on anything else; and the other is the "broad content", a function that yields the referent in every possible world, and depends on the external world. Narrow content is a conceptual role. Meaning needs both narrow and broad contents.

In the 1980s the US mathematicians Jon Barwise and John Perry proposed "situation semantics", a relation theory of meaning: the meaning of a sentence provides a constraint between the utterance and the described situation. Sentences stand for situations rather than for truth values. Properties and relations are primitive entities. Situations turn out to be more flexible than Kripke's possible worlds because they don't need to be coherent and don't need to be maximal. Just like mental states.

The US philosopher David Lewis even argued that possible worlds should be assumed to be real ("modal realism"). Things that may have been are no less real to him than things that actually are.

Truth-conditional Semantics

The US philosopher Donald Davidson was the main proponent of "truth-conditional semantics", which reduces a theory of meaning to a theory of truth. Tarski simply replaced the universal and intuitive notion of "truth" with an infinite series of rules which define truth in a language relative to truth in another language. Davidson would rather assume that the concept of "truth" need not be defined, that it is known to everybody. Then he can use the correspondence theory of truth to define meaning: the meaning of a sentence is defined as what it would be if the sentence were true. The task for a theory of meaning then becomes to generate all meta-sentences (or "T-sentences") for all sentences in the language through a recursive procedure.

This account of meaning relies exclusively on truth conditions. A sentence is meaningful in virtue of being true under certain conditions and not others. To know the meaning of a sentence is to know the conditions under which the sentence would be true.

A theory of a language must be able to assign a meaning to every possible sentence of the language. Just like Chomsky had to include a recursive procedure in order to explain the speaker's unlimited ability to "recognize" sentences of the language, so Davidson has to include a recursive procedure in order to explain the speaker's unlimited ability to "understand" sentences of the language.

Natural languages exhibit an additional difficulty over formal languages: natural languages contain "deictic" elements (demonstratives, personal pronouns, tenses) which cause the truth value to fluctuate in time and depend on the speaker. Davidson therefore proposed to employ a pair of arguments for his truth predicate, one specifying the speaker and one specifying the point in time.

In other words, Davidson assigns meanings to sentences of a natural language by associating the sentences with truth-theoretically interpreted formulas of a logical system (their "logical form").

The US philosopher William Lycan basically refined Davidson's meta-theory. Lycan's theory of linguistic meaning rests on truth conditions too.

All other aspects of semantics (verification conditions, use in language games, illocutionary force, etc.) are derived from that notion. A sentence is meaningful by virtue of being true under certain conditions and not others. However, instead of assigning only a pair of arguments to the truth predicate, Lycan defines truth as a pentadic relationship with the logical form, the context (truth is relative to a context of time and speaker, as specified by some assignment functions), the degree (languages are inherently vague, and sentences normally contain fuzzy terms and hedges) and the idiolect (the truth of a sentence is relative to the language of which it is a grammatical string).

Holism: Meaning is Relative
The US philosopher Willard Quine was the messiah of holism.

Quine's theory of "under-determination" originated in the sciences. Quine was profoundly influenced by an argument put forth by the French physicist Pierre Duhem: that hypotheses cannot be tested in isolation from the whole theoretical network in which they appear. Quine argued that an hypothesis is verified true or false only relative to background assumptions.

For every empirical datum there can be an infinite number of theories that explain it. Science simply picks the combination of hypotheses that seems more plausible. When a hypothesis fails, the scientist can always modify the other hypotheses to make it hold. There is no certain way to determine what has to be changed in a theory: any hypothesis can be retained as true or discarded as false by performing appropriate adjustments in the overall network of assumptions. No sentence has special epistemic properties that safeguard it from revision.

The so called Quine-Duhem thesis reads: no part of a scientific theory can be proved or disproved; only the whole can.

Ultimately, science is but self-conscious common sense.

Language is a special case. The empirical datum in this case is a discourse and the theory is its meaning. There are infinite interpretations of a discourse depending on the context. A single word has no meaning: its referent is "inscrutable". The meaning of language is not even in the mind of the speaker. It is a natural phenomenon related to the world of that speaker.

Quine thinks that the meaning of a statement is the method that can verify it empirically. But verification of a statement within a theory depends on the set of all other statements of the theory. Each statement in a theory partially determines the meaning of every other statement in the same theory.

In particular, the truth of a statement cannot be assessed as a function of the meaning of its words. An individual statement can be proved true or false only relative to the theory they belong to. Words do not have an absolute meaning. They have a meaning only relative to the other words they are connected to in the sentences that we assume to be true. The meaning of a sentence depends on the interpretation of the entire language. Its meaning can even change in time.

In general, the structure of concepts is determined by the positions that their constituents occupy in the "web of belief" of the individual.

In particular, it is impossible to define what a "correct" translation of a statement is from one language to another, because that depends on the interpretations of both entire languages. Translation from one language to another is indeterminate.

Meaning has no meaning. The only concept that makes sense for interpreting sentences is truth. A sentence can be true or false, but what it refers to is not meaningful.

Technically, Quine's ideas can be expressed in terms of variables. Values of variables cannot be fixed until the interpretation of the whole formal system is fixed (because of a famous theorem in mathematics, the Loewenheim-Skolem theorem). Thus his famous motto: "To be is to be the value of a variable".

The US philosopher Paul Churchland expanded Quine's holism by interpreting Quine's network of meanings as a space of semantic states, whose dimensions are all the observable properties. Each expression in the language is equivalent to defining the position of a concept within this space according to the properties that the concept exhibits in that expression. The semantic value of a word derives from its place in the network of the language as a whole.

Coherence

The correspondence theory of truth assumes that the definition of truth is in the world.

However, one can object that everything is ultimately in the mind and therefore the definition of truth is inside us. It is pointless to look for a definition in the world. Idealists (as opposed to materialists) believe this. This leads to a different theory of truth: truth can no longer be defined as the correspondence to the facts of the worlds, but has to be defined as the correspondence with the facts of the mind. The "coherence theory of truth" defines truth as coherence with the system of beliefs in one's mind: the statement "Snow is white" is true if the fact asserted by this statement this is coherent with all the other facts that are believed to be true.

Truth is defined by the set of coherent statements that make up a whole system of beliefs.

Any cosmological theory, for example, is of this kind: the truth of a statement about black holes cannot be verified (because we cannot travel into a black hole and not even get close to one) and therefore it only depends on whether it is coherent with the other "truths" of Physics.

Idealists believe that this is the only definition of truth that makes sense in general: we can never be sure of the world, therefore we can only assess whether a statement is coherent or not with our beliefs.

The US philosopher Charles Peirce pioneered the "pragmatist" approach to meaning that roughly says: the meaning of an idea consists in its practical effects on our daily lives. If two ideas have the same practical effects on us, they have the same meaning. He then defined accordingly: truth is the effect it has on us, and that effect is "consensus". Truth is not agreement with reality, it is agreement among humans. That agreement is reached after a process of scientific investigation. At the end of each such process, humans reach a consensus about what is "true" (e.g., that the Earth is not the center of the universe, that water is made up of hydrogen and oxygen, that Everest is the highest mountain on Earth).

Verificationism

Coherence put truth back into the mind of the observer. Verificationism put it back out into the world.

The British philosopher Michael Dummett criticized holism because it cannot explain how an individual can learn language. If the meaning of a sentence only exists in relationship to the entire system of sentences in the language, it would never be possible to learn it. For the same reason it should not be possible to understand the meaning of a theory, if its meaning is given by the entire theory and not by single components. Dummett's theory of meaning is instead a variant of intuitionistic logic: a statement can be said to be true only when it can be proved true in finite time (it can be "effectively decided", similar to the intuitionistic "justified").

The truth of a statement must be provable in a finite amount of time, otherwise the statement is not true. The statement "I will never win the Nobel prize" is provable (just wait until i die), but the statement "I am a genius" or "there will never be another like me" are not provable, and therefore their truth value cannot be determined. When we say that a statement is true, we mean that it can be verified. Dummett applies to the world at large the same rules that "intuitionists" applied to logic: to decide the truth of a statement is to prove a theorem. The proof determines truth. If no proof can be constructed, then there is no truth. Verification is not just a means to achieve truth: it is truth. The two concepts are virtually impossible to separate.

Similarly, the Finnish philosopher Jaakko Hintikka proposed a "game-theoretical semantics", whereby the semantic interpretation of a sentence is reduced to a game between two agents. The semantics searches for truth through a process of falsification and verification. The truth of an expression is determined through a set of domain-dependent rules which define a "game" between two agents: one agent is trying to validate the expression, the other one is trying to refute it. The expression is true if the truth agent wins. Unlike Dummett's verificationist semantics, Hintikka's is still a "truth-conditional" semantics.

The British psychologist Philip Johnson-Laird, too, believes that the meaning of a sentence is the way of verifying it. In his "procedural" semantics, a word's meaning is the set of conceptual elements that can contribute to build a mental procedure necessary to comprehend any sentence including that word. Those elements depend on the relations between the entity referred by that word and any other entity it can be related to. Rather than atoms of meanings, we are faced with "fields" of meaning, each including a number of concepts that are related to each other. The representation of the mental lexicon handles the intensional relations between words and their being organized in semantic fields.

Something is true if and only if its truth can be practically verified.

Externalism

The US philosopher Hilary Putnam attacked model-theoretic semantics from another perspective: in his opinion, it fails as a theory of meaning because meaning is not in the relationship between symbols and the world.

Putnam argued that "meaning is not in the mind". Putnam imagines a world called "Twin Earth" exactly like Earth in every respect except that the stuff which appears and behaves like water, and is actually called "water", on Twin Earth is a chemical compound XYZ. If one Earth and one Twin Earth inhabitants, identical in all respects, think about "water", they are thinking about two different things, while their mental states are absolutely identical. Putnam concludes that the content of a concept depends on the context. Meanings are not in the mind, they also depend on the objects that the mind is connected to.

Meaning exhibits an identity through time but not in its essence (such as the momentum, which is a different thing for Newton and Einstein but expresses the same concept). An individual's concepts are not scientific and depend on the environment. Most people know what gold is, and still they cannot explain what it is and even need a jeweler to assess whether something is really gold or a fake. Nonetheless, if some day we found out that Chemistry has erred in counting the electrons of the atom of gold, this would not change what it is. The meaning of the word "gold" is not its scientific definition, but the social meaning that a community has given it. It is not true that every individual has in her mind all the knowledge needed to understand the referent of a word. There is a subdivision of competence among human beings and the referent of a word is due to their cooperation.

Interlude: Intentionality

Intentionality (the ability of a system to refer to something outside it) is at the center of the philosophical debate on meaning.

Willard Quine's "instrumentalist theory" rejected intentionality altogether, considering it merely a linguistic trick. Quine thus denied psychology (that deals with desires, beliefs and hopes and so forth) the status of science.

Daniel Dennett reached a similar conclusion: the intentional stance is simply one of the possible ones, and it is just that, a stance. It is useful to attribute intentional states to other humans, and maybe even to non-human systems. Intentionality is only a feature of our language, a tool. The mind is not intentional because it does not exist, it is only a term in the vocabulary of folk psychology. The brain is not intentional because its states do not refer to anything: they simply are what they are, just like in any physical system.

Jerry Fodor's representational theory, instead, posits that the brain literally contains the intentional state that we ascribe to the mind. For Fodor, the vocabulary of folk psychology "is" a scientific language: it describes facts that are really happening in the brain, not just linguistic tools.

The US philosopher Ruth Millikan argued that intentionality is an objective, natural, biological feature of humans, that evolved over millions of years just like any other organ or limb of the human body. Intentionality is no more than the biology of belief, desire, hope and intention, which must be treated like any other biological object. Propositional attitudes (beliefs, hopes, desires, intentions) are biological devices, designed by evolution to have some effects on us. That effect is its content. Thus the effect has been determined by evolution. This strategy of treating intentionality as a biological feature can be extended to treating meaning as a biological phenomenon ("biosemantics"), because each sentence can be viewed as having a biological function (typically, helping us live in the world). "Sentences are basic intentional items". Intentionality is grounded in the relationship with the environment, a relationship determined by evolution. A mental content is the effect of a biological system designed by evolution to have that effect. Presumably, Millikan believes that desires and beliefs are physiological features of the brain, because they have evolved from generation to generation just like any other bodily organ has. On the other hand, the contents of these beliefs and desires are located outside the brain, and can be understood only by understanding their biological function, i.e. their evolutionary history. A belief is similar to the dance of a bee, a biological feature designed by evolution that refers to an object outside the head of the bee.

The "externalist theory" of the British philosopher Colin McGinn is an extension of Millikan's teleological theory. The mental content of humans is external to their minds for the simple reason that it was set by evolution and it refers to the environment. The cognitive life of humans has been shaped by evolution to cope with an external object, the environment. Beliefs and desires are brain states whose content is a relation to the external world.

The information-processing theory of the US philosopher Fred Dretske assumes that cognitive life consists in transforming analogue information (that comes from the sensory system) into digital information (that can become the processed as knowledge). For example, a smell per se is only a set of sensory data. Once it is analyzed and turned into the information that it corresponds to, the smell of a particular flower, we "know" a fact about the world. This transformation from analogue to digital is what creates a "belief". A belief is therefore a neural structure. This is what the system "believes". The content of that belief is thus defined by its informational origin and not by its behavioral effect. A concept is the link between the information origin and the behavioral effect. The semantic content becomes a cognitive content when it is transformed from a representational unit to a functional unit.

In the functionalist theory of the British philosopher Brian Loar beliefs and desires are real physical states with real causal powers, but they are wholly defined by and within the overall network of beliefs an desires. A belief is defined by its functional role within the network of the person's beliefs. In a sense, the mind "is" the network of prepositional attitudes.

The US philosopher John Searle thinks that intentionality can only be relative to a contextual "network" of other intentional states and to a "background" of pre-intentional stances. The background is necessary for the network to function. In other words, in order to perform a mental act, one must have a network of mental acts and they must be grounded in the real world. For example, I can "desire to see a film" only if I believe that the film exists and is showing. Thus the "network". And i can go and see the film only if I know how to drive the car and how to find the address. Thus the "background".

Deflationism

Of course, one can also claim that truth simply does not exist, and that is why it is so difficult to define.

In 1927 the British philosopher Frank Ramsey inaugurated "deflationary" thinking about truth by claiming that the word "true" is simply redundant: "It is true that the snow is white" does not say anything more than "The snow is white". By adding "It is true that" we are not adding anything: we are merely making it sound nicer.

Quine's "disquotationalism" follows from this claim: to ascribe truth to a statement merely means to remove the quotation marks.

For example, the statement "Snow is white" is true if and only if it is a fact that snow is white. Now remove the inessential words and what you have is: "Snow is white" is true if and only if snow is white. The truth predicate "is true" simply removes the quotation marks.

"Truth is disquotation". Quine concedes that the truth predicate (the expression "is true") has at least one useful function: it allows us to generalize, like when I state "everything I told you is true". By using the truth predicate, I can simplify what would otherwise be an infinite list of statements. But this is the only usefulness of the truth predicate: there is no need for a theory of truth, there is no nature of truth. The truth predicate is merely a linguistic expedient to generalize statements.

The problem remains, of course, that ordinary humans can easily grasp the concept of "true", whether Quine believes it to be a mere "disquotation" or not. There is something that we call "truth" in our minds.

Donald Davidson argues that truth is a primitive concept that cannot be defined via any other concept. In fact, no other concept would exist without the concept of truth.

Functionalist Truth

Ludwig Wittgenstein proposed a common-sense theory of truth: different statements can all be true without being true in the same way. This idea led him to "alethic pluralism", i.e. to accept that truth is a multi-faceted concept.

Likewise, the US philosopher Michael Lynch takes issue with the idea that there is "one" theory of truth. Lynch argues that there is a plurality of "truths", rather than a single all-encompassing theory of truth. For example, truth in ethics and truth in justice and truth in mathematics obey different laws. The nature of truth is difficult to find because truth doesn't have only one nature. One needs a different theory of truth for each domain, and that is precisely what ordinary humans employ in their daily lives.

Just like functionalism believes in "multiple realizations" of the same mental phenomenon, i.e. that the same mental state can be "realized" by different physical states (what matters being the function, not the "stuff"), Lynch believes that "truth" (a uniform concept across domains) can be realized by different theories in different domains.

For example, pain is a mental state that is causally related to some inputs (e.g., a sore finger), outputs (e.g., facial expression and sounds), and other mental states (e.g., unhappiness). Any state that realizes this causal role is called "pain", even if the pain due to a blister and the pain due to a cold are very different in nature.

Lynch claims that "truth" names a functional role, and that we all understand what that role is, regardless of what realizes it. Lynch compares this with the concept of "head of state": both the president of the United States, the king of Jordan, Fidel Castro and the chancellor of Germany are heads of state, although the way they got the job and the way they administer it vary greatly. The "function" of head of state, though, is understood the same way in the US, France and Cuba.

(A possible objection is that equality is sometimes merely a form of fuzziness: the closer you look, the less similar Castro and the king of Jordan are, and the less clear the term "head of state" is. One can suspect that "functional role" is a synonym for "vague definition". Relax the definition and just about anything in this universe will have the same "functional role" as anything else).

If truth is merely a functional role, if "to be true" is to play the "alethic" role, what exactly is that role?

Lynch thinks that truth is defined by an "alethic network", a set of interdependent definitions that, jointly, define each other: a proposition is whatever is true or false, a fact is what makes a proposition true or false, etc. Lynch claims that each "alethic concept" in the alethic network is defined by the role it plays in the network. One cannot grasp an alethic concept (truth, proposition, fact) without grasping them all. Each alethic concept depends on all of the others. Truth cannot be defined as "stand alone", but only as part of the broader definition of all alethic entities.

Truth is the property of playing the truth role in an alethic network.

There is one and only one concept of truth, but it can be realized in multiple ways.

Further Reading
Barwise, Jon & Perry, John: SITUATIONS AND ATTITUDES (MIT Press, 1983)
Barwise, Jon: THE SITUATION IN LOGIC (Cambridge Univ Press, 1988)

Castaneda, Hector-Neri: THINKING, LANGUAGE, EXPERIENCE (University of Minnesota Press, 1989)

Churchland, Paul: SCIENTIFIC REALISM AND THE PLASTICITY OF MIND (Cambridge Univ Press, 1979)

Davidson, Donald: INQUIRIES INTO TRUTH AND INTERPRETATION (Clarendon Press, 1984)

Dennett Daniel: KINDS OF MINDS (Basic, 1998)

Dretske, Fred: KNOWLEDGE AND THE FLOW OF INFORMATION (MIT Press, 1981)

Dretske Fred: EXPLAINING BEHAVIOR (MIT Press, 1988)

Dummett Michael: ELEMENTS OF INTUITIONISM (Oxford University Press, 1977)

Dummett, Michael: TRUTH AND OTHER ENIGMAS (Harvard Univ Press, 1978)

Dummett, Michael: SEAS OF LANGUAGE (Clarendon, 1993)

Fodor, Jerry: A THEORY OF CONTENT (MIT Press, 1990)

Hintikka, Jaakko: THE GAME OF LANGUAGE (Reidel, 1983)

Johnson-Laird, Philip: MENTAL MODELS (Harvard Univ Press, 1983)

Kirkham, Richard: THEORIES OF TRUTH (MIT Press, 1992)

Kripke, Saul: NAMING AND NECESSITY (Harvard University Press, 1972)

Lewis, David: "On The Plurality of Worlds" (Blackwell, 1986)

Loar, Brian: MIND AND MEANING (1981)

Lycan, William: LOGICAL FORM IN NATURAL LANGUAGE (MIT Press, 1984)

Lynch, Michael: THE NATURE OF TRUTH (MIT Press, 2001)

McGinn, Colin: MENTAL CONTENT (Blackwell, 1989)

Millikan, Ruth: LANGUAGE, THOUGHT AND OTHER BIOLOGICAL CATEGORIES (1984)

Moore, A.W.: MEANING AND REFERENCE (Oxford Univ Press, 1993)

Putnam, Hilary: MIND, LANGUAGE AND REALITY (Cambridge Univ Press, 1975)

Putnam, Hilary: REASON, TRUTH AND HISTORY (Cambridge Univ Press, 1981)

Putnam, Hilary: REPRESENTATION AND REALITY (MIT Press, 1988)

Quine, Willard: WORD AND OBJECT (MIT Press, 1960)

Quine, Willard: FROM A LOGICAL POINT OF VIEW (Harper & Row, 1961)

Quine, Willard: THE WEB OF BELIEF (Random House, 1978)

Quine, Willard: ONTOLOGICAL RELATIVITY (Columbia Univ Press, 1969)

Searle, John: INTENTIONALITY (Cambridge Univ Press, 1983)

Tarski, Alfred: LOGIC, SEMANTICS, METAMATHEMATICS (Clarendon, 1956)

A Timeline of Neuroscience

1590: Rudolph Goeckel's "Psychologia" introduces the word "psychology" for the discipline that studies the soul

1649: Pierre Gassendi's "Syntagma philosophiae Epicuri" argues that beasts have a cognitive life of their own, just inferior to humans

1664: Rene Descartes' "Treatise of Man" argues that the pienal gland is the main seat of consciousness (Great Minds Series):

1664: Thomas Willis' "Cerebral Anatomy" (1664) describes the different structures in the brain and coins the word "neurology"

1741: Emanuel Swedenborg's "The Economy of the Animate Kingdom" discusses cortical localization in the brain

1771: Luigi Galvani discovers that nerve cells are conductors of electricity

1796: Franz-Joseph Gall begins lecturing on phrenology, holding that mental faculties are localized in specific brain regions (of which 19 are shared with animals and 8 are exclusive to humans)

1824: Pierre Flourens' "Phrenology Examined" discredits Gall

1825: Jean-Baptiste Bouillaud's "Clinical and Physiological Treatise upon Encephalitis" describes patients who suffered brain lesions and lost their speech ability

1836: Marc Dax's "Lesions of the Left Half of the Brain Coincident With the Forgetting of the Signs of Thought" notes that aphasic patients (incapable of speaking) have sustained damage to the left side of the brain

1861: Paul Broca's "Loss of Speech, Chronic Softening and Partial Destruction of the Anterior Left Lobe of the Brain" single-handedly resurrects the theory of cortical localization of function

1865: Paul Broca's "Localization of Speech in the Third Left Frontal Convolution" suggests that the location of speech must be in the left hemisphere

1868: John Hughlings Jackson's "Notes on the Physiology and Pathology of the Nervous System" reports how damage to the right hemisphere impairs spatial abilities

1870: Eduard Hitzig and Gustav Fritsch discover the location of the motor fuctions in the brain

1873: Jean-Martin Charcot's "Lectures on the Diseases of the Nervous System" describes the neural origins of multiple sclerosis

1873: Camillo Golgi's "On the Structure of the Brain Grey Matter" describes the body of the nerve cell with a single axon and several dendrites

1874: Karl Wernicke determines that sensory aphasia (a loss of linguistic skills) is related to damage to the left temporal lobe

1874: Charles-Edouard Brown-Sequard's "Dual Character of the Brain" argues that education does not adequately target the right hemisphere

1876: John Hughlings Jackson discovers that loss of spatial skills is related to damage to the right hemisphere

1876: David Ferrier's "The Functions of the Brain" provides a map of brain regions specialized in motor, sensory and association functions

1890: Wilhelm His coins the word "dendrite"

1891: Santiago Ramon y Cajal proves that the nerve cell (the neuron) is the elementary unit of processing in the brain, receiving inputs from other neurons via the dendrites and sending its output to other neurons via the axon

1891: Wilhelm von Waldeyer coins the term "neuron" while discussing Santiago Ramon y Cajal's theory

1896: Albrecht von Kolliker coins the word "axon"

1897: Charles Sherrington coins the word "synapse"

1901: Charles Sherrington maps the motor cortex of apes

1903: Alfred Binet's "intelligent quotient" (IQ) test

1905: Keith Lucas demonstrates that below a certain threshold of stimulation a nerve does not respond to a stimulus and, once the threshold is reached, the nerve continues to respond by the same fixed amount no matter how strong the stimulus is

1906: Charles Sherrington's "The Integrative Action of the Nervous System" argues that the cerebral cortex is the center of integration for cognitive life

1911: Edward Thorndike's connectionism (the mind is a network of connections and learning occurs when elements are connected)

1921: Otto Loewi demonstrated chemical transmission of nerve impulses, proving that nerves can excite muscles via chemical reactions (notably acetylcholine) and not just electricity

1924: Hans Berger records electrical waves from the human brain, the first electroencephalograms

1924: Konstantin Bykov, performing split-brain experiments on dogs, discovers that severing the corpus callosum disables communications between the two brain hemispheres

1924: Hans Berger records electrical waves from the human brain, the first electroencephalograms

1925: Edgar Adrian shows that the message from one neuron to another neuron is conveyed by changes in the frequency of the discharge, the first clue on how sensory information might be coded in the neural system

1928: Otfried Foerster stimulates the brain of patients during surgery with electric probes

1933: Henry Dale coins the terms "adrenergic" and "cholinergic" to describe the nerves releasing the two fundamental classes of neurotransmitters, the adrenaline-like one and acetylcholine

1935: Wilder Penfield explains how to stimulate the brain of epileptic patients with electrical probes ("Epilepsy and Surgical Therapy")

1936: Jean Piaget's "The Origins of Intelligence in Children"

1940: Willian Van Wagenen performs "split brain" surgery to control epileptic seizures

1949: Donald Hebb's cell assemblies (selective strengthening or inhibition of synapses causes the brain to organize itself into regions of self-reinforcing neurons - the strength of a connection depends on how often it is used)

1951: Roger Sperry's "chemoaffinity theory" of synapse formation explains how the nervous system organizes itself during embryonic development via a genetically-determined chemical matching program

1952: Paul Maclean discovers the "limbic system"

1953: John Eccles' "The Neurophysiological Basis of Mind" describes excitatory and inhibitory potentials, the two fundamental changes that occur in neurons

1953: Roger Sperry and Ronald Meyers study the "split brain" of animals

1953: Eugene Aserinsky discovers "rapid eye movement" (REM) sleep that corresponds with periods of dreaming

1954: Rita Levi-Montalcini discover nerve-growth factors that help to develop the nervous system, thus proving Sperry's chemoaffinity theory

1957: Vernon Mountcastle discovers the modular organization of the brain (vertical columns)

1959: Michel Jouvet discovers that REM sleep originates in the pons

1962: Roger Sperry and Michael Gazzaniga discover that the two hemispheres of the human brain are specialized in different tasks

1962: David Kuhl invents SPECT (single photon emission computer tomography)

1964: John Young proposes a "selectionist" theory of the brain (learning is the result of the elimination of neural connections)

1964: Paul Maclean's triune brain: three layers, each layer corresponding to a different stage of evolution

1964: Lueder Deecke and Hans-Helmut Kornhuber discover an unconscious electrical phenomenon in the brain, the Bereitschaftpotential (readiness potential)

1964: Benjamin Libet discovers that the readiness potential precedes conscious awareness by about half a second

1968: Niels Jerne's selectionist model of the brain (mental life a continuous process of environmental selection of concepts in our brain - the environment selects our thoughts)

1972: Raymond Damadian builds the world's first Magnetic Resonance Imaging (MRI) machine

1973: Edward Hoffman and Michael Phelps create the first PET (positron emission tomography) scans that allow scientists to map brain function

1972: Jonathan Winson discovers a correlation between the theta rhythm of dreaming and long-term memory

1972: Godfrey Hounsfield and Allan Cormack invent computed tomography scanning or CAT-scanning

1977: Allan Hobson's theory of dreaming

1978: Gerald Edelman's theory of neuronal group selection or "Neural Darwinism"

1985: Michael Gazzaniga's "interpreter" (a module located in the left brain interprets the actions of the other modules and provides explanations for our behavior)

1989: Wolf Singer and Christof Koch discover that at, any given moment, very large number of neurons oscillate in synchrony and one pattern is amplified into a dominant 40 Hz oscillation (gamma synchronization)

1990: Seiji Ogawa's "functional MRI" measures brain activity based on blood flow

1994: Vilayanur Ramachandran proves the plasticity of the adult human brain

1996: Giacomo Rizzolatti discovers that the brain uses "mirror" neurons to represent what others are doing

1996: Rodolfo Llinas: Neurons are always active endlessly producing a repertory of possible actions, and the circumstances "select" which specific action is enacted

1997: Japan opens the Brain Science Institute near Tokyo

2009: The USA launches the Human Connectome Project to map the human brain

2012: Mark Mayford stores a mouse's memory of a familiar place on a microchip

2013: The European Union launches the Human Brain Project to computer-simulate the human brain

A Timeline of Artificial Intelligence

1935: Alonzo Church proves the undecidability of first order logic
1936: Alan Turing's Universal Machine ("On computable numbers, with an application to the Entscheidungsproblem")
1936: Alonzo Church's Lambda calculus
1941: Konrad Zuse's programmable electronic computer
1943: Warren McCulloch's and Walter Pitts' binary neuron ("A Logical Calculus of the Ideas Immanent in Nervous Activity")
1943: Kenneth Craik's "The Nature of Explanation"
1943: "Behavior, Purpose and Teleology" co-written by mathematician Norbert Wiener, physiologist Arturo Rosenblueth and engineer Julian Bigelow
1945: John Von Neumann designs a computer that holds its own instructions, the "stored-program architecture"
1946: The ENIAC, the first Turing-complete computer
1946: The first Macy Conference on Cybernetics
1947: John Von Neumann's self-reproducing automata
1948: Norbert Wiener's "Cybernetics"
1948: Alan Turing's "Intelligent Machinery"
1949: Leon Dostert founds Georgetown University's Institute of Languages and Linguistics
1949: William Grey-Walter's Elmer and Elsie robots
1950: Alan Turing's "Computing Machinery and Intelligence" (the "Turing Test")
1951: Karl Lashley's "The problem of serial order in behavior"
1951: Claude Shannon's maze-solving robots ("electronic rats")
1952: First International Conference on Machine Translation organized by Yehoshua Bar-Hillel
1952: Ross Ashby's "Design for a Brain"
1954: Demonstration of a machine-translation system by Leon Dostert's team at Georgetown University and Cuthbert Hurd's team at IBM, possibly the first non-numerical application of a digital computer
1956: Dartmouth conference on Artificial Intelligence
1956: Allen Newell and Herbert Simon demonstrate the "Logic Theorist"
1957: Newell & Simon's "General Problem Solver"
1957: Frank Rosenblatt's Perceptron
1957: Noam Chomsky's "Syntactic Structures" (transformational grammar)
1958: Oliver Selfridge's Pandemonium
1958: John McCarthy's LISP programming language
1959: John McCarthy's "Programs with Common Sense" (1949) focuses on knowledge representation
1959: John McCarthy and Marvin Minsky found the Artificial Intelligence Lab at the MIT
1959: Arthur Samuel's Checkers, the world's first self-learning program
1959: Yehoshua Bar-Hillel's "proof" that machine translation is impossible

1959: Noam Chomsky's review of a book by Skinner ends the domination of behaviorism and resurrects cognitivism
1960: Hilary Putnam's Computational Functionalism
1960: Bernard Widrow's and Ted Hoff's Adaline ((Adaptive Linear Neuron or later Adaptive Linear Element) that uses the Delta Rule for neural networks
1962: Joseph Engelberger deploys the industrial robot Unimate at General Motors
1963 John McCarthy moves to Stanford and founds the Stanford Artificial Intelligence Laboratory (SAIL)
1963 Irving John Good (Isidore Jacob Gudak) speculates about "ultraintelligent machines" (the "singularity")
1964: IBM's "Shoebox" for speech recognition
1965: Ed Feigenbaum's Dendral expert system
1965: Lotfi Zadeh's Fuzzy Logic
1966: Ross Quillian's semantic networks
1966: Joe Weizenbaum's Eliza
1967: Barbara Hayes-Roth's Hearsay speech recognition system
1967: Charles Fillmore's Case Frame Grammar
1967: Leonard Baum's team develops the Hidden Markov Model
1968: Glenn Shafer's and Stuart Dempster's "Theory of Evidence"
1968: Peter Toma founds Systran to commercialize machine-translation systems
1969: Marvin Minsky & Samuel Papert's "Perceptrons" kill neural networks
1969: First International Joint Conference on Artificial Intelligence (IJCAI) at Stanford
1969: Stanford Research Institute's Shakey the Robot
1969: Roger Schank's Conceptual Dependency Theory for natural language processing
1970: Albert Uttley's Informon for adaptive pattern recognition
1970: William Woods' Augmented Transition Network (ATN) for natural language processing
1971: Richard Fikes' and Nils Nilsson's STRIPS
1971: Ingo Rechenberg publishes his thesis "Evolution Strategies", a set of optimization methods for evolutionary computation
1972: Alain Colmerauer's PROLOG programming language
1972: Bruce Buchanan's MYCIN
1972: Terry Winograd's Shrdlu
1972: Hubert Dreyfus's "What Computers Can't Do"
1973: "Artificial Intelligence: A General Survey" by James Lighthill criticizes Artificial Intelligence for over-promising
1974: Marvin Minsky's Frame
1974: Paul Werbos' Backpropagation algorithm for neural networks
1975: Roger Schank's Script
1975: John Holland's genetic algorithms
1976: Doug Lenat's AM
1976: Richard Laing's paradigm of self-replication by self-inspection
1979: William Clancey's Guidon
1979: Cordell Green's system for automatic programming

1979: Drew McDermott's non-monotonic logic
1979: David Marr's theory of vision
1980: McCarthy's Circumscription
1980: Kunihiko Fukushima's Convolutional Neural Networks
1980: John Searle publishes the article ""Minds, Brains, and Programs" on the "Chinese Room" that attacks Artificial Intelligence
1980: John McDermott's Xcon
1980: Intellicorp, the first major start-up for Artificial Intelligence
1981: Danny Hillis' Connection Machine
1982: John Hopfield describes a new generation of neural networks, based on a simulation of annealing
1982: The Canadian Institute for Advanced Research (CIFAR) establishes Artificial Intelligence and Robotics as its very first program
1982: Japan's Fifth Generation Computer Systems project
1982: Teuvo Kohonen's Self-Organized Maps (SOM) for unsupervised learning
1982: Judea Pearl's "Bayesian networks"
1983: John Laird and Paul Rosenbloom's SOAR
1983: Geoffrey Hinton's and Terry Sejnowski's Boltzmann machine for unsupervised learning
1984: Valentino Breitenberg's "Vehicles"
1986: David Rumelhart's "Parallel Distributed Processing" rediscovers Werbos' backpropagation algorithm
1986: Paul Smolensky's Restricted Boltzmann machine
1987: Hinton moves to the Canadian Institute for Advanced Research (CIFAR)
1987: Chris Langton coins the term "Artificial Life"
1987: Stephen Grossberg's Adaptive Resonance Theory (ART) for unsupervised learning
1987: Marvin Minsky's "Society of Mind"
1987: Rodney Brooks' robots
1988: Hilary Putnam: "Has artificial intelligence taught us anything of importance about the mind?"
1988: Philip Agre builds the first "Heideggerian AI", Pengi, a system that plays the arcade videogame Pengo
1988: Fred Jelinek's team at IBM publishes "A statistical approach to language translation"
1990: Carver Mead describes a neuromorphic processor
1992: Thomas Ray develops "Tierra", a virtual world
1994: The first "Toward a Science of Consciousness" conference in Tucson, Arizona
1995: Geoffrey Hinton's Helmholtz machine
1996: David Field & Bruno Olshausen's sparse coding
1997: IBM's "Deep Blue" chess machine beats the world's chess champion, Garry Kasparov
1998: Yann LeCun's second generation Convolutional Neural Networks
1998: Two Stanford students, Larry Page and Russian-born Sergey Brin, launch the search engine Google

2000: Cynthia Breazeal's emotional robot, "Kismet"
2000: Seth Lloyd's "Ultimate physical limits to computation"
2001: Juyang Weng's "Autonomous mental development by robots and animals"
2001: Nikolaus Hansen introduces the evolution strategy called "Covariance Matrix Adaptation" (CMA) for numerical optimization of non-linear problems
2003: Hiroshi Ishiguro's Actroid, a robot that looks like a young woman
2003: Jackrit Suthakorn and Gregory Chirikjian at Johns Hopkins University build an autonomous self-replicating robot
2004: Mark Tilden's biomorphic robot Robosapien
2004: Ipke Wachsmuth's conversational agent "Max"
2005: Hod Lipson's "self-assembling machine" at Cornell University
2005: Honda's humanoid robot "Asimo"
2005: Andrew Ng at Stanford launches the STAIR project (Stanford Artificial Intelligence Robot)
2005: Boston Dynamics' quadruped robot "BigDog"
2006: Geoffrey Hinton's Deep Belief Networks (a fast learning algorithm for restricted Boltzmann machines)
2006: Osamu Hasegawa's Self-Organising Incremental Neural Network (SOINN), a self-replicating neural network for unsupervised learning
2007: Yeshua Bengio's Stacked Auto-Encoders
2008: Dharmendra Modha at IBM launches a project to build a neuromorphic processor
2008: Adrian Bowyer's 3D Printer builds a copy of itself at the University of Bath
2008: Cynthia Breazeal's team at the MIT's Media Lab unveils Nexi, the first mobile-dexterous-social (MDS) robot
2008: A 3D Printer builds a copy of itself at the University of Bath
2010: The New York stock market is shut down after algorithmic trading has wiped out a trillion dollars within a few seconds.
2010: Quoc Le's "Tiled Convolutional Networks"
2010: Lola Canamero's Nao, a robot that can show its emotions
2011: Nick D'Aloisio releases the summarizing tool Trimit (later Summly) for smartphones
2011: IBM's Watson debuts on a tv show
2011: Osamu Hasegawa's SOINN-based robot that learns functions it was not programmed to do
2012: Rodney Brooks' hand programmable robot "Baxter"
2013: John Romanishin, Kyle Gilpin and Daniela Rus' "M-blocks" at MIT
2014: Vladimir Veselov's and Eugene Demchenko's program Eugene Goostman, which simulates a 13-year-old Ukrainian boy, passes the Turing test at the Royal Society in London
2014: Li Fei-Fei's computer vision algorithm that can describe photos ("Deep Visual-Semantic Alignments for Generating Image Descriptions", 2014)
2014: Alex Graves, Greg Wayne and Ivo Danihelka publish a paper on "Neural Turing Machines"
2014: Jason Weston, Sumit Chopra and Antoine Bordes publish a paper on "Memory Networks"

Alphabetical Index of Names

Abelard, Pierre: 139
Aleksander, Igor: 208
Anderson, John: 48
Arbib, Michael: 60, 273
Armstrong, David: 24, 26
Ashby, Ross: 149
Austin, John: 285
Bacon, Francis: 139
Baddeley, Alan: 116
Baker, Mark: 227
Baldwin, JamesMark: 216
BarHillel, Yehoshua: 236
Barricelli, Nils: 166, 257
Bartlett, Frederic: 113
Barwise, Jon: 141, 295
Bates, Elizabeth: 241
Baum, Eric: 87
Bayes, Thomas: 133, 188, 206
Beardsley, Monroe: 277
Bekenstein, Jacob: 154
Bengio, Yeshua: 207
Berger, Hans: 76
Berger, Theodore: 101
Bergson, Henri: 117
Berkeley, George: 19
Berlin, Brent: 123
Bernstein, Nikolai: 147
Bickerton, Derek: 250
Black, Max: 273, 278
Block, Ned: 28, 30
Bonnet, Charles: 13
Boole, George: 140
Breal, Michel: 272
Brentano, Franz: 13, 33
Brillouin, Leon: 152
Broad, CharlesDunbar: 18
Broadbent, Donald: 115, 116
Brooks, Rodney: 171
Brouwer, Luitzen: 183
Brown, Jason: 95
Brown, Roger: 123, 241

Bruner, Jerome: 121, 159, 287
Buchanan, Bruce: 158
Buck, Ross: 74
Bykov, Konstantin: 74
Caianiello, Eduardo: 207
Cairn, Robert: 63
Campbell, Donald: 55
Cannon, Walter: 148
Carbonell, Jaime: 160
Chaitin, Gregory: 154
Chalmers, David: 18, 41
Changeux, JeanPierre: 86
Chapin, John: 101
Chomsky, Noam: 29, 49, 60, 88, 115, 126, 220, 221, 227, 239, 240, 296
Church, Alonzo: 144, 146
Churchland, Paul: 32, 211, 270, 298
Clark, Andy: 208
Cohen, Neal: 118
Conrad, Michael: 166
Craik, Kenneth: 47
Cytowic, Richard: 91
Damasio, Antonio: 89, 120
Darwin, Charles: 231, 245, 255
Davidson, Donald: 23, 25, 287, 296, 303
Davis, Ernest: 193
Dawkins, Richard: 15
Deacon, Terrence: 253
Deadwyler, Sam: 101
DeBiran, Maine: 117
DeJong, Jerry: 160
DeKleer, Johan: 195
Delgado, José: 101
DeMorgan, Augustus: 140
Dempster, Stuart: 188
Dennett, Daniel: 28, 30, 33, 34, 260, 301

DeSaussure, Ferdinand: 220, 269, 278
Descartes, Rene: 13, 25, 36, 239
Descartes, René: 12
Deutsch, David: 146, 167
Devlin, Keith: 234
Djourno, Andre: 101
Dobelle, William: 101
Donald, Merlin: 269, 270
Dowty, David: 232
Dretske, Fred: 34, 35, 38, 169, 302
Dreyfus, Hubert: 170, 172
Dubinsky, Donna: 206
Dubois, Didier: 191
Duhem, Pierre: 190, 297
Dummett, Michael: 299
Dunbar, Robin: 252
Dyson, Freeman: 155
Eccles, John: 16, 76
Edelman, Gerald: 83, 84, 227
Einstein, Albert: 20, 228, 300
Erikson, Eric: 132
Euler, Leonhard: 140
Eyries, Charles: 101
Fauconnier, Gilles: 52, 126, 239, 276, 280
Feigenbaum, Edward: 158
Feigl, Herbert: 21
Fetzer, James: 270
Feyerabend, Paul: 32
Fikes, Richard: 160
Fillmore, Charles: 232
Finke, Ronald: 52
Fisher, Ronald: 104
Fodor, Jerry: 28, 29, 33, 49, 53, 130, 208, 209, 226, 278, 295, 301
Forbus, Kenneth: 195
Freeman, Walter: 87, 104
Frege, Gottlob: 140, 142, 290, 295
Freud, Sigmund: 132, 138
Fukushima, Kunihiko: 207
Galvani, Luigi: 76
Gazdar, Gerald: 235
Gazzaniga, Michael: 73, 75, 88, 106

Gentzen, Gerhard: 291
George, Dileep: 206
Gerrold, David: 165
Geschwind, Norman: 95
Ginsberg, Matthew: 187
Goedel, Kurt: 141, 146, 154, 172
Goertzel, Ben: 87
Goldberg, David: 161
Goldenring, John: 106
Goldstein, Kurt: 69
Goodale, Melvyn: 55
Gopnik, Alison: 132
Gould, Stephen: 87, 245
Grady, Joe: 276
Greenfield, Patricia: 131
Gregory, Richard: 259
Grenander, Ulf: 206
Grice, Paul: 35, 272, 284, 289
Grossberg, Stephen: 207
Gupta, Alan: 184
Harnad, Stevan: 279
Hasegawa, Osamu: 207
Hawking, Stephen: 173
Hawkins, Jeff: 61, 206
Hayes, Pat: 193
HayesRoth, Barbara: 184
Haynes, John: 92
Hebb, Donald: 71, 82, 204
Heidegger, Martin: 36, 38, 171
Heisenberg, Werner: 154, 190
Helmholtz, Hermann: 46
Higuchi, Tetsuiya: 167
Hilbert, David: 140, 142, 143, 221
Hintikka, Jaakko: 300
Hinton, Geoffrey: 204, 205
Hjelmslev, Louis: 268
Hoffmeyer, Jesper: 269
Holland, John: 160, 161, 166
Holloway, Ralph: 253
Honderich, Ted: 38
Hopfield, John: 204
Hume, David: 13
Humphrey, Nicholas: 104, 251
Husserl, Edmund: 36, 170
Huxley, Thomas: 13

Indurkhya, Bipin: 279
Iriki, Atsushi: 103
Jablonski, Nina: 101
Jackendoff, Ray: 225, 233
Jackson, Frank: 24
Jackson, Hughlings: 75, 94
Jacobs, Lucia: 98
Jakobson, Roman: 218
James, William: 19, 68, 70, 71, 115, 117
Jeannerod, Marc: 56
Jerne, Niels: 83, 84
Jespersen, Otto: 248
Johnson, Mark: 62, 125
JohnsonLaird, Philip: 51, 53, 300
Joseph, Rhawn: 97, 254
Kant, Immanuel: 126, 274
KarmiloffSmith, Annette: 130, 210
Katz, Bernard: 76
Katz, Jerrold: 234
Keil, Frank: 126
Kelly, Kevin: 267
Kennedy, Philip: 101
Kittay, Eva: 278
Koehler, Wolfgang: 69
Kohonen, Teuvo: 207
Kolmogorov, Andrei: 154
Korzybski, Alfred: 219
Kosko, Bart: 190
Kosslyn, Stephen: 53
Kotelnikov, Vladimir: 153
Kripke, Saul: 24, 290, 293, 294
Kuipers, Benjamin: 195
Kupfer, Abraham: 84
Kurzweil, Ray: 265
Laird, John: 48
Lakoff, George: 52, 124, 125, 239, 274, 276
Landsteiner, Karl: 83
Langacker, Ronald: 125, 237, 238
Langer, Susanne: 246, 249
Langley, Pat: 161
Langton, Chris: 165
Lashley, Karl: 69, 213

LeCun, Yann: 207
Lee, TaiSing: 206
Leibniz, Gottfried: 19, 140
Lenat, Douglas: 161, 196
Lenneberg, Eric: 121, 226
Lewis, David: 24, 27, 296
Lieberman, Philip: 94, 249
Linnaeus, Carl: 228
Livanov, Mikhail: 80
Llinas, Rodolfo: 93
Loar, Brian: 302
Loewi, Otto: 76
Loritz, Donald: 241
Lucas, JohnRandolph: 172
Lucas, Keith: 76
Lyapunov, Aleksandr: 153
Lycan, William: 28, 31, 296
Lynch, Michael: 303
MacCormac, Earl: 277
Mach, Ernst: 19
MacLean, Paul: 96
Malsburg, Carl: 208
Mange, Daniel: 167
Marr, David: 29, 49, 50
Martin, James: 274
Martin, Per:-Lof 184
McCarthy, John: 155, 156, 186, 194
McCulloch, Warren: 202
McDermott, Drew: 187
McDermott, John: 158
McGinn, Colin: 53, 54, 302
Mehta, Mayank: 92
Meinong, Alexius: 33
MerleauPonty, Maurice: 62
Mettrie, Julien: 21
Micera, Silvestro: 102
Michalski, Ryszard: 159
Miller, Geoffrey: 104, 255
Miller, George: 115, 116, 261
Millikan, Ruth: 301
Milner, David: 55
Minsky, Marvin: 31, 57, 204
Mitchell, Tom: 159, 161
Mithen, Steven: 272
Montague, Richard: 235, 239, 290

Moore, Robert: 187
Mora, Francisco: 99
Morris, Charles: 268
Moscovitch, Morris: 120
Mountcastle, Vernon: 61, 85
Mueller, Anthonie: 99
Mumford, David: 206
Nagel, Thomas: 24
Nelson, Katherine: 213
Newell, Alan: 47, 48, 115
Newell, Allen: 55, 115, 158, 161, 177, 270
Newton, Isaac: 20, 140, 300
Nicolelis, Miguel: 102
Ockham, William: 139
Pattee, Howard: 166
Pavlov, Ivan: 68, 138, 207
Peano, Giuseppe: 140
Pearl, Judea: 205
Peirce, Charles: 60, 61, 140, 268, 299
Penrose, Roger: 16, 172, 173
Perry, John: 295
Piaget, Jean: 60, 127, 129, 130, 132, 224
Pinker, Steven: 226, 228, 230, 232
Pitts, Walter: 202
Place, Ullin: 23
Polya, George: 197
Popper, Karl: 14
Powers, William: 149
Prade, Henri: 191
Prieto, Luis: 269
Purves, Dale: 86
Putnam, Hilary: 28, 172, 290, 300
Pylyshyn, Zenon: 53
Quine, Willard: 33, 126, 297, 301
Ramachandran, Vilayanur: 73, 90, 245, 266
Ramon, Santiago: 76
Ramsey, Frank: 26, 303
Rao, Rajesh: 102
Ray, Thomas: 166
Reddy, Michael: 283
Reiter, Raymond: 187
Ribeiro, Sidarta: 92

Richards, Ivor: 273
Rizzolatti, Giacomo: 90
Robertson, Lynn: 75
Rorty, Richard: 32
Rosch, Eleanor: 122
Rosch, Eleanor: 122, 124, 125
Rosenblatt, Frank: 202
Rosenbloom, Paul: 48, 160
Ross, John: 225
Rucker, Rudy: 16, 173
Rumelhart, David: 205
Russell, Bertrand: 19, 20, 24, 25, 140, 141, 247, 295
Russell, Stuart: 174
Ryle, Gilbert: 22
Sadock, Jerrold: 285
Saffran, Jenny: 206
Sapir, Edward: 213, 248, 276
Savage, Leonard: 188
Scaruffi, Piero: 1, 2, 7, 140, 183, 284, 290, 294
Schacter, Daniel: 117, 119, 120
Schank, Roger: 58, 61, 233
Searle, John: 17, 18, 23, 34, 39, 169, 174, 286, 302
Sebeok, Thomas: 269
Sejnowsky, Terrence: 204
Selz, Otto: 56
Semon, Richard: 114
Shafer, Glenn: 188
Shannon, Claude: 35, 153
SheetsJohnstone, Maxine: 62
Shyreswood, William: 139
Simon, Herbert: 47, 55, 115, 158, 161, 177, 179, 270
Skinner, Burrhus: 68
Sloman, Aaron: 172
Smart, John: 23
Smith, Linda: 63
Smolensky, Paul: 205
Sperry, Roger: 73, 75, 81
Spinoza, Baruch: 18
Stevens, Anthony: 97
Stich, Stephen: 28, 30
Stocco, Andrea: 102
StuartMill, John: 140, 294

Szilard, Leo: 152
Tarski, Alfred: 143, 293, 296
Taylor, Timothy: 263
Thelen, Esther: 63
Thom, Rene: 269, 270
Thorndike, Edward: 70
Tipler, Frank: 155, 166
Tolman, Edward: 114
Tomasello, Michael: 134
Trubetzkoy, Nicholas: 218
Tulving, Endel: 117, 119
Turing, Alan: 144, 145, 164, 168, 173, 174, 176
Ullman, Shimon: 50
VonHumboldt, Alexander: 213
VonNeumann, John: 144, 164, 166
Vygotsky, Lev: 128, 129, 214, 288
Wallace, Alfred: 149
Watson, John: 21
Weaver, Warren: 35, 153
Wernicke, Karl: 75
Wertheimer, Max: 69
Wheeler, John: 16
Wheelwright, Philip: 277
Whitehead, Alfred: 19
Whorf, Benjamin: 213, 276
Wiener, Norbert: 138, 148
Wierzbicka, Anna: 236
Wilson, Deirdre: 286
Wilson, Frank: 249
Wilson, Neil: 287
Winograd, Terry: 171, 172
Winson, Jonathan: 120
Wittgenstein, Ludwig: 121, 124, 283, 291, 303
Young, John: 81, 99
Zadeh, Lofti: 191
Zadeh, Lotfi: 124, 189
Zeki, Semir: 86

www.ingramcontent.com/pod-product-compliance
Lightning Source LLC
Chambersburg PA
CBHW051626170526
45167CB00001B/78